FM 5-20H

WAR DEPARTMENT FIELD MANUAL

CAMOUFLAGE MATERIALS AND MANUFACTURING TECHNIQUES

1944

WORLD WAR II

CIVILIAN REFERENCE EDITION

UNABRIDGED HISTORIC ARMY HANDBOOK ON
PROCURING CONCEALMENT ITEMS AND SUPPLIES

U.S. WAR DEPARTMENT

Doublebit Press

www.doublebitpress.com | Cherry, IL, USA

Original content under the public domain; unrestricted for civilian distribution.
Originally published in 1944 by the U.S. War Department.

A Part of the Military Manual Series

Doublebit Press Civilian Reference Edition ISBNs
Hardcover: 978-1643892207
Paperback: 978-1643892214

WAR DEPARTMENT FIELD MANUAL
FM 5-20H

CAMOUFLAGE MATERIALS AND MANUFACTURING TECHNIQUES

WAR DEPARTMENT • JULY 1944

United States Government Printing Office
Washington : 1944

WAR DEPARTMENT,
WASHINGTON 25, D. C. 1 July 1944.

FM 5-20H, Corps of Engineers Field Manual, Camouflage Materials and Manufacturing Techniques, is published for the information and guidance of all concerned.

[A.G. 300.7 (11 Mar 44).]

BY ORDER OF THE SECRETARY OF WAR:

G. C. MARSHALL,
Chief of Staff.

OFFICIAL:
J. A. ULIO,
Major General,
The Adjutant General.

DISTRIBUTION:

As prescribed in paragraph 9a, FM 21-6: except D(5), OCE (500) Armd Force Sch (300), Comd & Gen Stf Sch (300), Engr Sch (500), FA Sch (300), Inf Sch (300), TD Sch (300), AAF Cam Sch #1 (200) Cav Sch (200); RTC (200); B(5); R(4); Bn(4); C 1-4, 8-11, 19, 35, 44, 55(2); C 5-7, 17, 18 (4); I Bn 5 (100); I C 5 (10).

I Bn 5: T/O&E 5-95, Engr Cam Bn; 5-465, Engr Cam Bn, Avn.

I C 5: T/O&E 5-97, Engr Cam Co., Sep.

(For explanations of symbols, see FM 21-6.)

CONTENTS

This manual supersedes paragraphs 8-13, FM 5-20, 1 June 1940; TM 5-269, 7 October 1942; and TM 5-268, 24 June 1941.

	PARAGRAPHS	PAGE
CHAPTER 1. PURPOSE AND SCOPE	1-3	5
CHAPTER 2. ARTIFICIAL MATERIALS		
Section I. Paints, Stains, and Dyes	4-6	6
Paints	5	6
Tables I-X		8
Other materials and methods for coloring surfaces	6	16
Tables XI-XXVI		18
Section II. Adhesives	7-9	28
Application of adhesives	8	29
Tables XXVII-XXIX		30
Field-expedient adhesives	9	32
Tables XXX-XXXII		34
Section III. Texturing Materials	10-12	37
Preparation	11	38
Application	12	39
Table XXXIII		40
Section IV. Screening Materials	13-17	44
Cloth and cloth substitutes	14	44
Small-mesh nets	15	44
Tables XXXIV-XXXV		45
Field-expedient mat screens	16	48
Other screening materials	17	48
Section V. Supporting Materials	18-19	50
Large-mesh netting	18	50
Tables XXXVI-XXXVII		51
Repair of twine nets	19	56

	PARAGRAPHS	PAGE
Section VI. Garnishing Materials	20-23	65
Cloth	21	65
Tables XXXVIII-XXXIX		65
Metal	22	71
Other garnishing materials	23	72
Section VII. Structural Materials	24-28	76
Wire, cable, and field expedients	25	76
Tables XL-XLI		77
Fibrous supporting materials	26	79
Other materials	27	79
Table XLII		79
Posts, stakes, and other structural materials	28	80
CHAPTER 3. NATURAL MATERIALS		
Section I. General	29-34	81
Section II. Ground Patterns and Natural Materials	35-38	82
Two-dimensional ground patterning	37	83
Spreading material	37*a*	83
Scarification	37*b*	84
Spot and pattern irrigation	37*c*	85
Burning dry grass	37*d*	85
Grass planting	37*e*	86
Mowing	37*f*	90
Fertilizing	37*g*	92
Chemical killing agents	37*h*	94
Three-dimensional ground patterning	38	95
Section III. Overhead Cover and Natural Materials	39-45	96
Natural materials on artificial supports	40	96
Choosing cut foliage	41	98
Foliage preservatives	42	99
Planting vines	43	100
Tables XLIII-XLVI		101
Transplanting small trees and shrubs	44	106
Transplanting large trees	45	112
Table XLVII		113
CHAPTER 4. TOOLS AND EQUIPMENT		
Section I. Simple Tools, Heavy Equipment, and Painting Equipment	46-55	115
Section II. Miscellaneous Equipment	56-58	120
Section III. Field-expedient Manufacturing Devices	59-60	122
Table XLVIII		135

	PARAGRAPHS	PAGE
CHAPTER 5. PREFABRICATED ISSUE MATERIALS		
Section I. Camouflage Clothing	61-62	138
Section II. Net Sets	63-71	139
CHAPTER 6. USES OF CLAY, PLASTER, PAPIER MACHE, AND SIMILAR PLASTIC MATERIALS		
Section I. General	72	140
Section II. Making Walls	73	141
Section III. Methods of Moulding	74	142
Section IV. Methods of Simulating Terrain Features	75	146
CHAPTER 7. COVERAGE TABLES		
Notes on Coverage tables	76	147
Tables XLIX-LXIII		148
CHAPTER 8. SPECIFICATIONS, CAMOUFLAGE MATERIALS AND EQUIPMENT		
Section I. Specifications Listed by Issuing Authority	78-82	162
Section II. Specifications Listed by Subject	83-94	163
INDEX		166

CHAPTER 1

PURPOSE AND SCOPE

1. GENERAL.—This manual is designed as a source of quick reference data on camouflage materials and field manufacturing techniques. It is, however, beyond the scope of this manual to indicate what methods and materials are to be used in any given situation. A knowledge of the basic principles of camouflage (FM 5-20, Camouflage, Basic Principles) combined with a knowledge of techniques for concealing particular materiel or installations (FM 5-20, A to H) will enable the using officer to select from the materials available those suitable to the solution of his problem.

2. SCOPE.—This manual is intended to serve troops both in the zone of the interior and in theaters of operations. It describes issue materials and equipment and also many substitutes which can be manufactured from locally available material. The compilation of all reported field-expedient materials and devices has not been attempted, but at least one of each basic type is included. No references are included as to costs because these vary considerably.

3. ORGANIZATION OF MANUAL.—The Table of Contents gives the basic organization of this manual. However, if quick reference to one aspect of a particular subject is desired, the reader may find the cross-reference index, page 166, useful and time-saving. The index is organized alphabetically by main subject headings with subsidiary headings for specific topics under each main heading when appropriate. Cross references are given where the need for them is apparent. References to "issue" items, unless otherwise specified, mean Corps of Engineers issue materials and equipment.

FIGURE 1.—A factory for the production of colored garnishing strips in the field.

CHAPTER 2

ARTIFICIAL MATERIALS

		PARAGRAPHS
Section I.	Paints, Stains, and Dyes	4–6
II.	Adhesives	7–9
III.	Texturing Materials	10–12
IV.	Screening Materials	13–17
V.	Supporting Materials	18–19
VI.	Garnishing Materials	20–23
VII.	Structural Materials	24–28

SECTION I. PAINTS, STAINS, AND DYES

4. GENERAL.—Commercial paints and stains are the most familiar materials used for coloring surfaces. However, other materials may be used to change the color of a surface—for example, crankcase oil drainings, crude molasses, and native colored earths in the form of mud. The choice of a coloring material for a particular job is determined by:

(1) Color desired.

(2) Surface on which it is to be applied.

(3) Finish desired—mat or glossy.

(4) Climatic conditions—temperature, rainfall, and the amount of brilliant sunlight, which influences the fading rate.

(5) Availability.

(6) Durability required.

5. PAINTS.—*a.* In combat zones, the choice of paints is determined by availability. Normally, issue paints are available; but if not, supplies may be procurable from the depots of Allied Forces or captured enemy supplies; local products commercially manufactured may be used; or, if necessary, materials may be manufactured by field-expedient methods.

b. For large-scale work in rear areas, the cheapest paint, per gallon, is not always the least expensive; maintenance and durability must be considered.

c. To obtain the best results, paints must be used correctly under climatic and weather conditions for which they are suitable. Paint failure is usually caused by improper mixing, unfavorable conditions, improper application, or by a paint film which is not suited to the surface to which it is applied.

d. Normally, the requirements of camouflage paint are that it have a dull finish, be non-fading and infrared reflectant, and be able to cover in one coat. Most camouflage paints obtain the quality of infrared reflectance from the natural earth pigments used to give them their color. Green camouflage paints, however, obtain their infrared quality from chromium oxide. Because of the important strategic value of chromium it is essential to avoid waste or

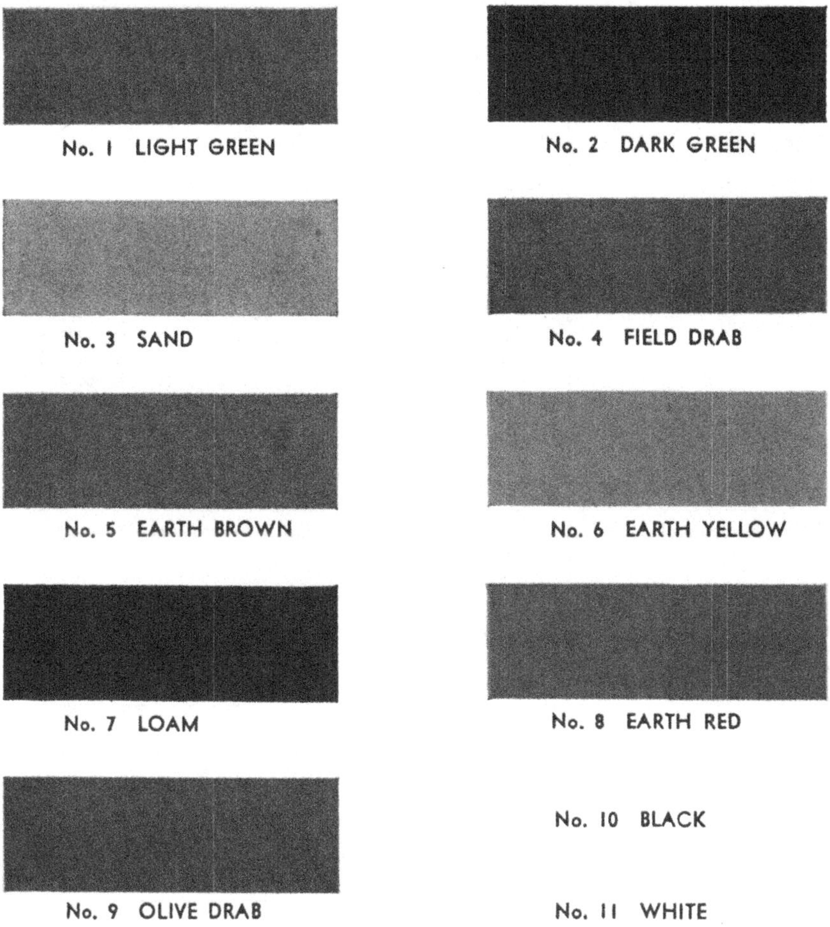

FIGURE 2.—The standard camouflage colors.

misuse of green camouflage paints. Their use is justified only where it is vital that green painted surfaces should photograph on infrared film with the same value as adjacent vegetation. Ordinary green paints, when so photographed, appear dark, whereas infrared-reflecting camouflage paints appear light, as does natural deciduous vegetation.

e. Most issue and commercial paints are procurable in the 11 standard camouflage colors (fig. 2). In spite of color standardization, there is considerable variation in hue between lots and between the products of different manufacturers.

f. To give various surfaces a white coating for blending with snow terrain, white camouflage paint is normally used. Whitewash, alabastine, or calcimine mixed with water may be used as substitutes when temperatures permit the use of water.

g. For detailed information on camouflage paints, see tables I–XXVI; chapter 7 discusses paint coverage.

Table I.—Ready-mixed, oil-type camouflage paint.

Types issued	Ready-mixed in 1- and 5-gallon cans.
Colors available	Standard camouflage colors.
Sources and stock number	Issue: Light green .... 52-5472.55 Dark green .... 52-5472.3 Sand .... 52-5472.75 Field drab .... 52-5472.5 Earth brown .... 52-5472.35 Earth yellow .... 52-5472.45 Loam .... 52-5472.6 Earth red .... 52-5472.4 Olive drab .... 52-5472.65 Black. 52-5472.2 White .... 52-5472.79 Commercial.
Recommended uses	Wood. Primer (table XI) should be used when a more permanent coating is desired. Primed metal. Rubber.
Durability	Good—excellent protection from weathering.
How prepared	Miscible with mineral spirits or gasoline.
How applied	Brushing. Spraying. Thin with 1 quart of thinner to 1 gallon of paint.
Drying time	About 24 hours.
Resultant surface	Flat finish.
How removed	Not practicable.
Climatic limitations	Not suitable for use in freezing temperatures. In cold climates dries slowly.
Specification	Corps of Engineers Tentative Specification No. T-1215.
Coverage	See chapter 7.

Table II.—Oleoresinous paint (emulsifiable).	
Types issued	Paste, in 5-gallon cans.
Colors available	Standard camouflage colors
Source and stock number	Issue: Light green .... 52–5395.55 Dark green .... 52–5395.3 Sand .... 52–5395.75 Field drab .... 52–5395.5 Earth brown .... 52–5395.35 Earth yellow .... 52–5395.45 Loam .... 52–5395.6 Earth red .... 52–5395.4 Olive drab .... 52–5395.65 Black .... 52–5395.2 White .... 52–5395.79 Commercial.
Recommended uses	Fabrics, primarily. Primed metal. Primed wood. Rubber.
Durability	Good—about 9 to 12 months; does not wash off or chalk or fade excessively. Not as durable as oil paint.
How prepared	Miscible with water or organic solvents such as kerosene, gasoline, mineral spirits, turpentine, and naphtha. For fabrics: 1 gallon of paint is reduced with 3 gallons of solvent. For other surfaces: 1 gallon of paint is reduced with 1 gallon of solvent. Paste or mixed paint must *not* be left in opened cans or volatile emulsifier will evaporate, making paint incapable of emulsifying.
How applied	Dipping. Spraying. Brushing.
Drying time	On fabrics: 15 minutes to 2 hours, depending on temperature and humidity. On wood and metal: Not over 6 hours.
Resultant surface	Flat and lusterless.
How removed	Not practicable.
Climatic limitations	Paste will not freeze (stable at temperatures from —40° F. to 160° F.). When reduced with gasoline or naphtha, it may be used in sub-zero temperatures. Low temperatures do not prevent its drying, although it takes longer. May be applied under cold, damp conditions. Impervious to rain once paint film is dry.
Specification	Corps of Engineers Tentative Specification No. T–1279.
Coverage	See chapter 7.

Table III.—Protein-binder, cold-water camouflage paint.

Types available	Powder, usually issued in 5-pound pasteboard containers. Paste, available in 5-gallon containers weighing about 50 pounds.
Colors available	Standard camouflage colors.
Source and stock number	Issue (powder type only): Light green.............................. 52–5241.55 Dark green.............................. 52–5241.3 Sand.............................. 52–5241.75 Field drab.............................. 52–5241.5 Earth brown.............................. 52–5241.35 Earth yellow.............................. 52–5241.45 Loam.............................. 52–5241.6 Earth red.............................. 52–5241.4 Olive drab.............................. 52–5241.65 Black.............................. 52–5241.2 White.............................. 52–5241.79 Commercially in paste or powder form.
Recommended uses	Fabrics—paste form has better penetration. Other surfaces *only* when need for color is temporary.
Durability	Poor—except under very favorable weather conditions, only slightly water-resistant, not washable, chalks and lightens in color, and in sunlight forms a brittle film.
How prepared	Miscible with water. POWDER. Reduce in proportions of 10 pounds of powder to 1 gallon of water. Must be used immediately after mixing. PASTE. One gallon of paste is mixed with ½ gallon of water to make about 1½ gallons of paint.
How applied	Dipping. Spraying. Brushing.
Drying time	Must set for 48 hours before it becomes even slightly rain-resistant. Dries to touch in 30 minutes to 2 hours and dries to handle within 6 hours.
Resultant surface	Hard, flat surface.
How removed	Scrubbing with scrub brush soaked in water and ammonia or soap.
Climatic limitations	Paste form must be protected from freezing while in storage, but can be used after thawing. Neither form can be used in freezing temperatures. Paste type in storage under conditions of heat is likely to decompose and become unusable.
Specification	Corps of Engineers Tentative Specification No. T–1093.
Coverage	See chapter 7.

NOTE: This paint is stocked at present, but its procurement is discontinued

TABLE IV.—Gasoline-soluble camouflage paint.	
Types issued	Powder, in 5-gallon cans (contents 20 pounds; 1 pound = 1 quart).
Colors available	Olive drab only is stocked in engineer depots. Commercially available in standard camouflage colors.
Source and stock number	Issue: Olive drab............................ 52-5318.65 Commercial.
Recommended uses	Previously coated surfaces—for application of non-permanent coatings, such as identification marks or patterns on vehicles, guns, aircraft, and equipment. Paint is readily removable with gasoline. Blackout purposes. Not suitable for canvas, unpainted wood, or metal.
Durability	Impermanent—durable for approximately 2 months.
How prepared	Miscible with gasoline or mineral spirits to consistency desired. Should be stirred vigorously. After gasoline is added, 1 to 2 hours are required to dissolve powder in solvent before paint is ready for use. Nine pounds of powder to 1 gallon of solvent make 1½ gallons of paint. For field measuring, one 5-gallon can of powder may be mixed with 2 gallons of solvent. Do not use near open flame or in a closed room because of fumes and inflammable nature of this paint.
How applied	Brushing. Spraying.
Drying time	15 to 60 minutes.
Resultant surface	Flat.
How removed	Scrub brush or rag soaked in gasoline.
Climatic limitations	Excellent for use in freezing temperatures.
Specification	Corps of Engineers Tentative Specification No. T-1227.
Coverage	See chapter 7.

NOTE: This material will also be stocked in paste form, which makes paint easier to mix and to remove.

TABLE V.—Lusterless enamel.

Types issued	Ready-mixed in 1-gallon cans.
Colors available	Olive drab and black.
Source and stock number	Issue: Olive drab..... 52–3415.65–5 Black. 52–3415.2–3
Recommended uses	Primed metal surfaces.
Durability	Excellent.
How prepared	Ready for application. May be thinned with mineral spirits.
How applied	Brushing. Spraying. Reduce with 15 parts by volume of thinner to 85 parts by volume of paint.
Drying time	Quick drying.
Resultant surface	Flat.
How removed	Not practicable.
Climatic limitations	Can be used in sub-zero temperature, but will dry slower and will require additional thinner.
Specification	Corps of Engineers Tentative Specification No. T–1599.
Coverage	See chapter 7.

Table VI.—Camouflage pigmented bituminous emulsions.

Types issued	Liquid, in 55-gallon drums.
Colors available	Olive drab only is stocked in engineer depots. All standard camouflage colors except white are available commercially.
Source and stock number	Issue: Olive drab.......................... 52-5160.65-55 Commercial
Recommended uses	Concrete. A prime coat of non-pigmented bituminous emulsion is recommended. Asphalt or tarred roofs. Bituminous surfaces. Granules—woodchips, sawdust. Damp surfaces. Do not use on machinery or apply to moving parts.
Durability	Fair—chalks rapidly, fades; the black bleeds through, especially in hot weather. Does not provide an impermeable film.
How prepared	Miscible with water in proportion of approximately 2 gallons of water to 1 gallon of paint. If used over metal, metal must first be coated with primer to prevent corrosion.
How applied	Brushing. Spraying Asphalt distributor.
Drying time	Will not be harmed by rain once film has dried. Film dries to touch in 4 hours and remains elastic throughout life of coating.
Resultant surface	Low gloss.
How removed	May be removed by burning: apply layer of fine sand 2 to 3 inches deep; saturate with gasoline; ignite; remove residue with stiff brushes or scrapers.
Climatic limitations	If allowed to freeze, paint is unfit for use, even if warmed, because it separates and cannot be re-emulsified. A special type is made commercially which withstands freezing. Suitable for use under humid conditions and on damp surfaces. In cold weather the film may crawl. Softens and melts in hot weather.
Specification	Corps of Engineers Tentative Specification No. T-1224, Class B.
Coverage	See chapter 7.

NOTE: Surfaces coated with bituminous emulsions cannot later be painted with any other type of paint; the use of these materials should be restricted to surfaces for which it is specifically recommended.

Table VII.—Camouflage cement water paints.	
Types available	Type I. Powder, cement. Type II. Protein, powder, modified cement. Type III. Oil paste, modified cement.
Colors available	Standard camouflage colors. Range of light colors usually used because cement acts as both pigment and binder. Darker colors are not as durable.
Source and stock number	Commercial.
Recommended uses	Powder types. If not subject to abrasion, use on: Tile. Brick. Masonry. Concrete. Stucco. Plaster, except gypsum plaster. NOTE: Not recommended for wood, bituminous surfaces, or for runways, roads, or other traffic-bearing surfaces. Paste type. Galvanized iron. Primed metal.
Durability	Good—does not powder, chip, or rub off.
How prepared	Miscible with water in equal volumes.
How applied	Brushing. May be used with some types of heavy-duty spray equipment.
Drying time	30 minutes to 6 hours.
Resultant surface	Hard, opaque, flat, lusterless.
How removed	Not practicable.
Climatic limitations	Cannot be applied in freezing temperatures.
Specification	Engineer Board Procurement No. 245.
Coverage	See chapter 7.

Table VIII.—Camouflage oil-type runway paint.	
Types available	Ready-mixed.
Colors available	Standard camouflage colors.
Source and stock number	Commercial.
Recommended uses	Concrete and bituminous surfaces exposed to abrasive effects of traffic. Asbestos. Wood and wood chips. Composition tiles.
Durability	Excellent—but dependent on amount ot wear.

TABLE VIII.—Camouflage oil-type runway paint. *Continued.*	
How prepared	Ready for application. If necessary, can be thinned either with mineral spirits or turpentine—1 gallon paint to ¼ gallon gasoline or V.M. & P. naphtha.
How applied	Spraying. Brushing. Asphalt distributor.
Drying time	Less than one hour.
Resultant surface	Flat.
How removed	Not practicable.
Climatic limitations	Should not be applied to wet or damp surface.
Specification	Engineer Board Procurement No. 244.
Coverage	See chapter 7.

TABLE IX.—Camouflage bituminous-emulsion adhesive.	
Types issued	Ready-mixed in 55-gallon drums.
Colors available	Dries to a blackish-brown.
Source and stock number	Issue: 52–1308.5–5 Commercial.
Recommended uses	Stabilized earth. Concrete. Surfaces previously coated with bituminous materials. Earth, cinders, gravel. Prime coat for pigmented bituminous emulsion. May be used on damp surfaces.
Durability	Impermanent—but since it is inexpensive it may be desirable to recoat rather than use a more expensive material initially. For ground painting, a priming coat is recommended.
How prepared	Miscible with water. To prepare, add four times its volume of water. If a wetting agent (2 percent by weight of tetra sodium pyrophosphate) is added, better penetration results.
How applied	Spraying. Brushing. Asphalt distributor.
Drying time	Dry to touch in 8 hours.
Resultant surface	Medium gloss, which dulls rapidly on exposure.
How removed	By burning. Apply layer of fine sand 2 to 3 inches deep over surface to be removed; saturate with gasoline; ignite; remove residue with stiff brushes or scrapers.
Climatic Limitations	Freezing breaks down the emulsion and renders the material useless. May crawl or wrinkle in cold weather.
Specification	Corps of Engineers Tentative Specification No. T–1224, Class A.
Coverage	See chapter 7.

Table X.—Cutback asphalt.	
Types issued	Type I. Slow cure (S.C.) Type II. Medium Cure (M.C.). Type III. Rapid Cure (R.C.), in 55-gallon drums.
Colors available	Dark brown to black.
Source and stock number	Issue: R.C.. . . . . .59-1400.6-55 Commercial.
Recommended uses	S.C. for dirt, cinders, gravel; has deep penetration. M.C. for concrete and dirt. R.C. for concrete and dirt. When applied to concrete, surface must be absolutely dry.
Durability	Good—dependent on growth of vegetation through coating when applied on earth.
How prepared	TYPE I (S.C.) is thinned with diesel oil. TYPE II (M.C.) is thinned with kerosene. TYPE III (R.C.) is thinned with gasoline.
How applied	Asphalt distributor. Spraying. Brushing.
Drying time	Variable, but quick drying.
Resultant surface	Mat.
How removed	By burning. Apply layer of fine sand 2 to 3 inches deep over surfaces to be removed; saturate with gasoline; ignite; remove residue with stiff brushes or scrapers.
Climatic limitations	None.
Specification	Federal specification No. SS-A-671.
Coverage	See chapter 7.
Substitute	Coal tar may be used as a substitute for cutback asphalt.

6. OTHER MATERIALS AND METHODS FOR COLORING SURFACES.—In addition to the paints listed in the foregoing tables, other materials and methods may be used for coloring surfaces (tables XIII to XXVI).

a. Primers.—Priming coats are normally necessary or desirable only on surfaces not previously coated. The principal use of primers (table XI) is to increase the covering power of camouflage paints and thus reduce cost, and in the case of metal surfaces to provide an impermeable film to prevent corrosion. Their use also facilitates the adhesion of the paint coat.

b. Stains and dyes.—(1) Stains (tables XII and XVI) are of less value than paints because they lack durability, provide little protection from weathering and abrasion, and have less stability of color.

(2) The issue fabric dye and canvas-refinishing compound (tables

XVII and XVIII) are practical for the use for which they are intended. Native dyes, such as indigo, cochineal, and gamboge, have little durability and color permanence, and their use is not recommended. Certain natural materials may be used to produce dyes to color fabric if issue or commercial paints are unavailable. A red dye may be made from the inner bark of birch trees, and from sorrel roots. A blue dye may be made from elderberries, from the roots of yellow iris, and from the berries of the privet with the addition of alum and salt. A yellow dye may be made from the inner bark of crab-apple or ash trees, from the stem and roots of the barberry, and from the leaves of birch trees. A green dye may be made from the bark of the elder, from the bark of larch trees with the addition of alum, and from the leaves of the lily of the valley. A brown dye may be made from the bark of oak, birch, or alder trees, from the roots of walnut trees, and in the autumn from pine needles. A black dye may be made from alder bark, from the roots of the blackberry with the addition of salts of iron, and from the bark of the elder with the addition of coppers.

c. Chemicals.—The use of chemicals, which are not items of issue, is limited to altering the color of natural live vegetation, primarily grass turf. The two types, fertilizers and killing agents, are discussed in detail in paragraphs 37*g* and 37*h*.

d. Field-expedient paints.—To make paints (tables XIX-XXIII) in the field from local materials requires ingenuity, but there are few localities where suitable earth pigments are not available. In civilized regions, local paint manufacturers may be employed. In other regions, native labor and materials may be used. Suitable binding mediums may be known and used by local natives, and it will pay to investigate native practices. In West Africa, for example, copal, a resinous gum, is available and is made into a varnish by the natives. It dries quickly, makes a good bond, and when combined with local colored earth pigments makes a satisfactory paint.

(1) The following procedure may be a guide in the manufacture of field-expedient paints:

(*a*) Search for source of various colored earths, rocks, or clays.

(*b*) Determine binding medium to be used.

(*c*) Select site for factory with regard to location of pigments, raw material for binder, water, availability of transportation, equipment, and labor.

(*d*) Obtain or build necessary processing equipment.

(2) The range of pigments depends on the variety of local clays and rocks. The principal pigment not usually available in nature is black. Two methods for making black pigment are described in the paragraph below. Normally, local earths must be processed before use to obtain a uniform fine texture. Colored earths from large ant hills, in regions where they are common, have been found to provide better and more easily ground pigment than other sources in the same region. Earths with a heavy sand content do not make good pigments. Colored earths are first dried, then pulverized by grinding in a mortar with a pestle. When earth is thoroughly pulverized, the finest pigment is separated through a sieve or screen, or by flotation. Equipment for this purpose is described in paragraph 60*h* (3), (4), and (5). Rocks are broken up, crushed, and pulverized, and the pigment separated in a similar way.

(3) *How to make a black pigment.*—Used coffee grounds are placed in a bucket or suitable container and heated over a fire. Eighteen pounds of

Table XI.—Oil-type camouflage paint primer for wood.	
Types issued	Ready-mixed in 1-quart containers.
Colors available	Dull red.
Source and stock number	Issue: 52-6724.2-5 Commercial.
Recommended Uses	Primarily as a primer for ready-mixed, oil-type camouflage paint applied to hard or absorbent wood surfaces, either vertical or horizontal.
Durability	Good.
How prepared	Stir with paddle.
How applied	Brushing. Spraying.
Drying time	Dries hard enough to recoat within 24 hours (50° to 90° F.).
Climatic limitations	Not suitable for sub-freezing temperatures.
Specification	Corps of Engineers Tentative Specification No. T-1287.

Table XII.—Camouflage face paint.	
Types issued	Stick form. Divided into two camouflage colors in cylindrical metal container. It contains an insect repellent which is effective from 4 to 6 hours. It is non-toxic.
Colors available	Light green and loam for white troops. Light green and sand for colored troops. White and loam for troops operating in snow.
Recommended Uses	Toning down and patterning face and hands of soldier for protective concealment in combat.
Durability	Good—not affected by perspiration, will not rub off from contact with brush and grass, or wash off by immersion in water.
How prepared	Ready for use.
How applied	Rub stick over face and hands.
Resultant surface	Dull, opaque film.
How removed	With soap and water.
Climatic limitations	Suitable for use at temperatures of 20° to 150° F.
Specification	Engineer Board Procurement No. 234.

NOTE: This material will be an item of future issue. Procurement, storage, and issue may be a responsibility of the Quartermaster Corps.

TABLE XIII.—Stains for wood surfaces.	
Types issued	Type I. Oil stains, ready-mixed. Type II. Asphalt stains, ready-mixed. Type III. Creosote stains, ready-mixed.
Colors available	Non-standard color range; variable with manufacturer.
Source and stock number	Commercial.
Recommended uses	New wood. Previously stained surfaces.
Durability	Fair—fair color premanence. Type III has a limited preservative effect on wood.
How prepared	Ready for application, but may be thinned with mineral spirits or turpentine.
How applied	Brushing. Spraying. Asphalt distributor.
Drying time	About 8 to 12 hours.
Resultant surface	Flat. Type I may be painted over with any other type of coating. Type II cannot be painted over as the asphalt will bleed through.
How removed	Not removable.
Climatic limitations	None.
Specifications	No government specification available.
Coverage	See chapter 7.

coffee grounds produce a pound of pigment. The fire should have a low flame to prevent ignition of volatile smoke; in case of ignition, a cover is placed over container to snuff out flame. Occasional stirring is necessary to obtain homogeneous charring and to prevent oxidation of the carbon formed. Mixing or stirring is increased as heating progresses; in the last stage continuous stirring is maintained. When the thick smoke stops, it signifies the volatile matter has been removed and the grounds are thoroughly carbonized. This carbon must be transferred to a closed container to allow the material to cool without the carbon oxidizing. It is then ground by any suitable means and the resulting powder sifted through a fine screen. The finer the powder, the better the paint.

(4) A black pigment may be obtained by burning crankcase oil.

e. Coloring by application of texture.—The application of naturally colored or precolored texturing materials is another method for coloring surfaces. Usually granules are affixed to a surface by an adhesive (par. 7).

f. In many cases our troops may use various camouflage paints from supplies of Allied Nations. Data on some of these paints are given in tables XXIV-XXVI.

TABLE XIV.—Stains for concrete.

Types issued	Type I. Xylene and Tar (RT-2). Type II. Kerosene and Asphalt.
Colors available	Brownish black is only serviceable color.
Source and stock number	Issue: Xylol-Xylene .... 51-9036.5-5 Commercial.
Recommended uses	Concrete. Tile roofs.
Durability	Only brownish black is of sufficient permanence for exterior use.
How prepared	Type I. Prepared with 20 percent by volume of xylene or any volatile coal tar solvent and 80 percent by volume of tar (RT-2). Increasing proportion of xylene makes color a lighter brown. Type II. Prepared with 75 percent by volume of kerosene to 25 percent by volume of asphalt.
How applied	Brushing. Asphalt distributor. Spraying.
Drying time	Fairly rapid.
Resultant surface	The surface coating of type II stain remains tacky and spotted or mottled in appearance and has a medium gloss, which disappears in a few days.
How removed	Not practicable.
Climatic limitations	None.
Specification	No government specification available.
Coverage	Data not available.

TABLE XV.—Field-expedient tentage stain No. 1.	
Type issued	Field expedient.
Colors available	Dark gray.
Source and stock number	Manufacture by troops.
Recommended uses	Tentage.
Durability	Lasted for a period of 7 weeks in Bengal in spite of heavy rain.
How prepared	BILL OF MATERIALS: Local earth. Soot. G.I. or field-expedient soap. Paraffin. Water. PROCEDURE: Mix local earth with soap solution in proportion of ½ pound of soap to 8 gallons of water in which soot soaked in a little paraffin has been added.
How applied	Brushing. Swabbing with a rag.
Drying time	Variable, but fairly rapid.
Resultant surface	Flat, lusterless.
How removed	Not practicable.
Climatic limitations	Not recommended for freezing temperatures.

TABLE XVI.—Field-expedient tentage stain No. 2.	
Type issued	Field expedient.
Colors available	Dark gray.
Source and stock number	Manufacture by troops.
Recommended uses	For faded or undyed tentage.
Durability	Good. Rain washes off stain to some extent.
How prepared	BILL OF MATERIALS: Charcoal. Water. PROCEDURE: Obtain charcoal by burning scrap lumber and crushing the residue as fine as possible. Mix this charcoal with water to desired tone.
How applied	Swab on with a rag.
Climatic limitations	Not recommended for freezing temperatures.

Table XVII.—Fabric dye.	
Type issued	Dye, in powder form. Size of container not available.
Colors available	Army drab D Q for dyeing white fabrics olive drab. Army green Q for dyeing white fabrics green. Army olive D for dyeing khaki fabrics green. Army olive H Q for changing OD No. 8 to OD No. 7.
Source and stock number	Quartermaster Corps.
Recommended uses	Dyeing or re-dyeing underwear, handkerchiefs, socks, towels, mosquito bars, uniforms, and bandoleers.
Durability	Limited fastness—prolonged exposure to weather, or repeated laundering will fade the color.
How prepared	BILL OF MATERIALS: Dye. Common salt (substitute: sea water). Water. Mess-kit spoon. Scale. Pail. Standard *metal* laundry washer (provided temperatures of at least 160° F. can be obtained; wooden laundry washers must *not* be used). See factory method (par. 60*e*). PROCEDURE: For 20 pounds of dry articles to be dyed, proportions are: 24 gallons of water, 2 pounds of salt, 1½ ounces of dye (2 rounded mess-kit spoonfuls of dye *a* or *b*, or 1 spoonful of dye *c* or *d*). Weigh dry articles to be dyed. Limit to be placed in laundry washer is approximately ½ its rated capacity. Add hot water (about 120° F.) to prescribed level. Add salt and run for 5 minutes. Dissolve dye in a pail of nearly boiling *fresh* water, stirring well to obtain complete solution. Slowly add to the turning laundry wheel. Turn on steam and raise temperature at least to 160° F. for 15 to 20 minutes. Turn off steam and operate for 15 to 20 minutes. Drain water. Rinse twice for 3 to 5 minutes in cold water. Remove articles and dry. After dyeing, the laundry wheel must be thoroughly cleaned with hot water and bleach solution and rinsed before being used for washing clothes. *Procedure in open kettle:* G.I. can (32 gallons) placed over a trench fire. Fill ¾ full of water and heat to 120° F. Add 2 pounds of salt. Add 20 pounds of articles and stir for 3 to 5 minutes. Dissolve dye as described above and add *slowly,* stirring constantly. Continue heating and stirring for 30 minutes at near boiling point. Rinse and dry.

Table XVIII.—Canvas refinishing compound.	
Types issued	Type I. Water, weather, and mildew resistant; ready-mixed in 1-gallon cans. Type II. Water, weather, mildew, and fire resistant; ready-mixed in metal drums.
Colors available	Color A. Dark olive drab (olive drab No. 7). Color B. Olive drab. Color C. Green. (Any other color may be requisitioned from Quartermaster Corps by a force commander.)
Source and stock number	Issue: Quartermaster Corps: Color A 51C-15180 Color B 51C-1585 Color C stock number not available. Corps of Engineers: Color A 52-2740.65-01.
Recommended uses	Fabrics—for coloring or re-coloring: Tentage. Duck. Canvas. Clothing.
Durability	Good.
How prepared	By stirring. Gasoline may be used as a diluent if needed.
How applied	Place on flat surface and apply by: Rubbing. Brushing with stiff paint or scrubbing brush. Spraying is not satisfactory. As compound is inflammable, do not apply in presence of flames or cigarettes.
Drying time	At least 2 hours are required before article is ready for use.
Resultant surface	Mat finish.
How removed	Not practicable.
Climatic limitations	Suitable for all climates.
Specifications	Quartermaster Corps Tentative Specification, O.Q.M.G. No. 3.

Table XIX.—Field-expedient emulsified paint.

Types issued	Field expedient—2 types.
Colors available	Dependent upon range of earth colors procurable.
Source and stock number	Manufacture by troops.
Recommended uses	Dust layer for earth-surfaced runways. Fabrics. Wood, but only where rainfall and abrasion are at a mi i-mum. Concrete: fairly satisfactory. Metal: unsatisfactory resistance to rain and abrasion.
Durability	Satisfactory—for short periods. Fair durability of color and film properties on fabric; poor adhesion and abrasion resistance on wood and metal.
How prepared	TYPE I. BILL OF MATERIALS: Crankcase oil—salvage or waste. Ground clay or powdered carbon. Gasoline, for thinning. Water. Pigmented soil. PROCEDURE: two gallons of water are mixed gradually with 1 gallon of used crankcase oil and ½ to ¼ gallon of ground clay. Screened pigmented soil is added to give desired color. This mixture is thinned to the right consistency with gasoline or mineral spirits. TYPE II. BILL OF MATERIALS: Crankcase oil—salvage or waste. Clay or diatomaceous earth. G.I. soap or field-expedient soap (table XX). Water. Pigmented soil. PROCEDURE: one and one-half bars of G.I. soap or an equivalent quantity of field-expedient soap (table XX) are dissolved in 3 gallons of water. Resultant solution is added gradually to 1 gallon of used crankcase oil, stirring constantly. One gallon of ground clay is then slowly stirred in and enough screened pigmented soil is added to give desired color.
How applied	Dipping—for fabric. Brushing—not very satisfactory. Spraying—not very satisfactory.
Drying time	About 1 to 5 hours—fairly rapid on fabrics, slowly on metals.
Resultant surface	Glossy on metal. Mat on other surfaces.
How removed	Not practicable.
Climatic limitations	Not usable in freezing temperatures; slower drying in cold weather.
Coverage	On runways, 100 square feet per gallon. Other data not available.

Table XX.—Field-expedient soap (for field-expedient paint).	
Type issued	Field expedient.
Source and stock number	Manufacture by troops.
Recommended uses	To be used as substitute for G. I. soap in manufacture of field-expedient paint as described in table XIX.
How prepared	BILL OF MATERIALS: Waste grease from mess kitchen. Wood ashes. PROCEDURE: Burn hard wood and collect the ashes. Make a sieve from a large tin can by punching holes in the bottom. Place ashes in bottom of this can and pour water into it. Collect the water passing through, which is, in fact, a lye. Heating waste grease and lye together will result in a soap.
How used	This soap is dissolved in water and used to emulsify used crankcase oil to make an emulsified oil as described in table XIX.

Table XXI.—Raw molasses as a field-expedient paint.	
Type issue	Field expedient.
Colors available	Brown.
Source and stock number	Commercial sugar refineries; practicable only in areas where crude molasses is plentiful.
Recommended uses	Lay dust. Tone earth, as on runway. In extreme emergency, as an adhesive.
Durability	Unstable. Not waterproof.
How prepared	Use as supplied. Thin with water; the more water, the less adhesion.
How applied	Asphalt distributor. Brushed with broom. Mobile spray equipment. May be covered with thin coating of local earth.
Drying time	Never dries; always sticky.
Coverage	5 to 20 square feet per gallon. Complete data on results of field use are not available.

TABLE XXII.—Field-expedient cowdung paint.

Type issued	Field expedient.
Colors obtainable	Medium drab to black.
Source and stock number	Manufactured by troops.
Recommended uses	Porous surfaces, as: Asbestos sheeting. Brick work. Plaster.
Durability	Good—in Bengal; weathering qualities elsewhere are unknown.
How prepared	BILL OF MATERIALS: Fresh unscreened cowdung. Coal tar. Slaked lime (dry) or cement. Water. PROCEDURE: Place cowdung in drums and add just sufficient water to cover. (To 5 lbs. of cowdung add 2 lbs. of water.) Allow to stand for 3 days. Mix thoroughly. Add slaked lime or cement to coal tar, then mix thoroughly with soaked cowdung. The approximate ratio by *volume* for different tones is: *TONE* — *SOAKED COWDUNG* — *COAL TAR* — *LIME* Medium drab — 40 — 2 — 1 Dark drab — 40 — 4 — 2 Very dark drab — 40 — 6 — 3 Black — 40 — 8 — 4
How applied	Surfaces should be well wetted at time of application. Apply liberally to cover thoroughly.
Coverage	Variable. Normally 450 square feet per 100 pounds of mixed paint.

TABLE XXIII.—Used crankcase oil as field-expedient paint.

Type issued	Salvage.
Colors available	Brownish.
Source and stock number	Motor pools. Air bases.
Recommended uses	To tone down concrete. To lay dust. To tone down stabilized soil. To tone down spoil around entrenchments.
Durability	Fair.
How prepared	Ready for application. May be mixed with native colored earths by hand or in motor-driven mixer.
How applied	Spraying. Asphalt distributor.
Climatic limitations	May be applied under any weather conditions.
Coverage	See chapter 7.

Table XXIV.—British standard camouflage paints.

Standard types stocked in Royal Engineer Depots:

Wool-grease emulsion. Bituminous emulsion. Flat oil paint.

British Standard Camouflage Color No.	Color Equivalent	British Standard Camouflage Color No.	Color Equivalent
1	Dark brown	8	Grass green
1A	Brown-black	9	Light green
2	Medium brown	10	Dark red
3	Light brown	11	Light red
4	Coffee	11A	Tile red
5	Light gray	11B	Desert sand
6	Dark green	12	Medium gray
6A	Green-black	13	Dark gray
7	Medium-green	14	Black

Wool-grease emulsion paint suitable for application on:

Asbestos.	Concrete (not roads).	Wood.
Brick.	Corrugated iron.	Fabric.

Stocked in colors 1A, 2, 7, 11A, 14.
Water repellent, but will not resist rust.

Bituminous emulsion suitable for application on:

Asbestos.	Concrete.	Fabric.
Brick.	Corrugated iron.	

Stocked in colors 1A, 2, 7, 11A, 14.

Flat Oil Paint—"Paint Prepared For Use"—suitable for application on:

Brick, weathered. Wood Corrugated iron, weathered. Fabric.

Stocked in a wider range of colors than the other two paints.

Reference: (British) Military Training Pamphlet No. 46, Part III, 1942.

Table XXV.—Camemulsion.

Type issued	Paste made locally. Issued in 56-pound bags for the British Army in the Middle East.
Colors available	Cream, Chocolate, Sand, Light green, Dark green, Black, Buff, Terracotta, Pink
Source and stock number	British Army, Royal Engineers.
Recommended uses	Fabric. Unplaned and planed wood.
Durability	Poor—unless well rubbed in.
How prepared	Miscible with water: slowly add 3 parts water to 1 part paste; stir continuously; knead. Stir while using.
How applied	Sprinkling; then rubbing in until color appears on other side of fabric. Dipping (small articles). Rubbing with mop.
Drying time	Not known.
Coverage	2,000 square feet of fabric per 56-pound bag.

Table XXVI.—Camcolour.			
Type issued	Powder. Locally manufactured in 56-pound bags for the British Army in the Middle East.		
Colors available	Milk Cream Sand Black	Buff Terracotta Light green	Pink Chocolate Dark green
Source and stock number	British Army, Royal Engineers.		
Recommended uses	Galvanized iron. Bituminous surfaces.	Primed metal. Wood	
Durability	When hardened, is insoluble in water, but does not protect unprimed metals from rain or dampness.		
How prepared	Miscible with water or oil. For non-absorbent surfaces, mix 2 parts of powder with 1 part, by volume, of water. Resultant mix will not run when applied to a vertical surface. Mix must be continually stirred while in use. For absorbent surfaces, mix 3 parts of powder with 2 parts of water, by volume; wet surface thoroughly with water and apply. For planed wood, mix 1 part powder with 5 parts of linseed oil by volume.		
How applied	Hand application. Spraying. Note: Film should not exceed ⅛ inch in thickness.	Brushing. Mopping.	
Drying time	About 12 hours.		
Resultant surface	Mat.		
Climatic limitations	Cannot be applied in freezing temperatures. Cannot be applied in a rain.		
Coverage	About 1,500 square feet per 56-pound bag.		

SECTION II. ADHESIVES

7. GENERAL.—*a. Uses.*—In camouflage, adhesives serve to adhere texturing materials to various surfaces. They may be applied to most surfaces. They fix texturing materials to surfaces such as:

(1) *Traffic-bearing surfaces,* such as roadways, runways, and taxiways. In this case the adhesive must form a firm, durable bond because the granules also form a wearing surface.

(2) *Non-traffic-bearing surfaces,* such as roofs, vehicles, windows, and sides of buildings. Less durable adhesives serve the purpose, depending on the climate, length of expected service, and specific granules to be used. Some special adhesives, which have a limited usefulness, include ethyl cellulose, varnish, and ordinary glue. These are used for small-scale jobs, such as coating window glass for reduction of light reflectance or blackout control.

b. Sources of adhesives.—Adhesives may be obtained by issue; by purchasing commercial adhesives; and by the manufacture of field-expedient adhesives by troops, when natural raw materials are available and the necessary equipment can be improvised.

c. Selection of adhesives.—A number of factors influence the selection of an adhesive for the solution of a camouflage problem. These factors are:

(1) Whether surface is traffic-bearing or non-traffic-bearing.

(2) *Size of granules.*—Small granules require a dense, light adhesive such as bituminous emulsion. Large-size granules require adhesives such as asphalt cement.

(3) *Climate.*—The climatic limitations of the different adhesives are included in the tables of adhesives (tables XXVII-XXXII).

d. Characteristics of issue adhesives.—Adhesive coats are asphalt cement, cutback asphalt, or bituminous emulsion (tables XXVII-XXIX). The only distinction between these materials is the form in which the asphalt is applied to the surface. After the asphalt cement has cooled and after the petroleum solvent or water has evaporated from the cutback or emulsion, the remaining asphalt is identical in all three cases.

(1) Asphalt cement is preferable under ordinary circumstances.

(2) Asphalt, paving grade, with a penetration within the limits of 50 to 100 is the best adhesive for use in the United States. For tropical use, asphalt with a penetration within the limits 40 to 60 is recommended. For uses in cold climates, asphalt with a penetration between 85 to 100 is suggested.

(3) Although bituminous emulsion can be applied to damp surfaces, there is danger of a poor bond in the presence of excess dampness. The emulsion will not furnish a satisfactory bond until the original dampness of the surface and the water in the emulsion have dried completely. This requires more time than would be required by permitting only the surface to dry and then applying asphalt cement. In tropical climates where high humidity exists and frequent rains occur, the emulsions have proved unsatisfactory because of the time required for the surface to dry thoroughly.

8. APPLICATION OF ADHESIVES.—Application of adhesives is a simple resurfacing operation. On large areas the best practices of highway construction are followed. On any surface, however, do not attempt adhesion of the surface coat when either the surface or the texturing material is wet. If possible, apply adhesives when temperature is high for the locality. The surface and the texturing materials must be free of dust, oil, or other foreign matter.

a. Preparation of surface.—(1) Wash all types of large-scale surfaces with brush and water. Be sure surface is dry before applying adhesive. Smaller surfaces may require only the removal of dust. Dust, oil, and foreign matter prevent the formation of a bond between the adhesive and the surface and must be removed.

(2) *Apply prime coat* on concrete surfaces. There is no necessity for a prime coat on clean bituminous surfaces.

(*a*) The prime coat consists of a thin asphaltic liquid. Upon application to a non-bituminous surface, the liquid will not only coat the surface with a thin asphaltic film, but will also tend to penetrate and seal it; this is particularly important for work done at low temperatures.

(*b*) The prime coat recommended is R.C.-O, a rapid-curing solution, thinned with about 50 percent white gasoline naphtha. Apply 24 hours in advance of adhesive coat. The rate of application depends somewhat on the skill of the operator, but should be .05 to .1 gallons per square yard, accord-

Table XXVII.—Bituminous emulsion, class A.	
Availability	Source and stock number: Issue: 55-gallon drums; 52–1308.5–5. Commercial. Specification number: Corps of Engineers Tentative Specification No. T–1224, Class A.
Preparation	Ready-mixed.
Application	Surfaces on which applied: Concrete. Bituminous.
How applied	Asphalt distributor. Spraying. Brushing with broom for small areas.
Coverage	.1 to .4 gallons per square yard.
Climatic limitations	Lowest atmospheric temperature for application, 32° F. Unsatisfactory in tropical climates.
Durability	Good—poor bonding in presence of excess water.
Time before surface can be used	48 hours, depending on weather conditions.
Types of granules suggested	Medium-hard asphalt chips, ¼ inch to ½ inch. Wood chips. Other similar granules.
Time of distribution of granules over this adhesive	10 minutes after application of adhesive.

Table XXVIII.—Asphalt cement.	
Availability	Source: Commercial. Specification number: Federal Specification SS–A–706 (type AP–6).
Preparation	Ready-mixed.
Application	Surface on which applied: Concrete. Bituminous.
How applied	Asphalt distributor. Brushing.
Coverage	.1 to .3 gallon per square yard.
Climatic limitations	Lowest atmospheric temperature for application, 45° F.

TABLE XXVIII.—Asphalt cement. *Continued.*

Durability	Good.
Time before surface can be used	30 minutes.
Type of granules suggested	Large size: Asphalt chips. Tan bark. Crushed rock. Wood chips. Coarse gravel.
Time of distribution of granules over this adhesive	Immediately after application of adhesive.

TABLE XXIX.—Cutback asphalt (R.C. Grade 50-100 Penetration).

Availability	Source and stock number: Issue: 55-gallon drums; 59-1400.6-55. Commercial. Specification number: Federal specification SS-A-671.
Preparation	Ready-mixed.
Application	Surfaces on which applied: Concrete. Bituminous.
How applied	Asphalt distributor. Spraying by mobile spray unit. Brushing on with broom for small areas.
Coverage	.1 to .4 gallon per square yard.
Climatic limitations	Application possible at sub-freezing temperatures when the thinner is gasoline.
Durability	Good.
Time before surface can be used.	24 hours.
Type of granules suggested	Wood chips. Bagasse. Other similar small materials. Fine gravel. Slate granules.
Time of distribution of granules over this adhesive.	10 minutes after application of adhesive.

FIGURE 3.—A taxiway being coated with adhesive by heavy-duty, hand-operated spray equipment.

ing to the absorptive characteristics of the surface; the heavier application is appropriate for porous surfaces.

b. Methods of application.—(1) *On large areas,* such as roadways, runways, and taxiways, concrete and bituminous surfaces are treated alike. Asphalt is preferred as the adhesive coat, but tar—grade RT-10—may be used. Apply either adhesive with an asphalt distributor or with heavy-duty spray equipment, as in figure 3.

(2) *On small areas,* vertical or horizontal, apply adhesive with heavy-duty equipment, by hand-brushing with brooms, or with a swab.

9. FIELD-EXPEDIENT ADHESIVES.—*a.* In deciding upon an adhesive for emergency use, the following points are considered:

(1) Availability of solvents or extractants and manufacturing equipment required for their preparation.

(2) Adhesive qualities of product on type of surface to which it is to be applied.

(3) Bonding strength of product for texturing material to be used.

(4) Effect of applied material upon color and other appearance factors.

(5) Durability under expected weather conditions.

(6) Ease of removal to facilitate changes of camouflage treatment, if required.

(7) Time and manpower available.

b. Field-expedient adhesives and sources for adhesives are described below and in tables XXX-XXXII.

(1) Raw molasses as supplied by commercial sugar refineries may be

FIGURE 4.—Field-expedient adhesive being prepared from green vegetation.

used in areas where cane molasses is available in large quantities. It is very unstable and not waterproof. It may be applied by an asphalt distributor or by mobile spray equipment, or it may be brushed on with a broom. Use water to thin out or to remove.

(2) Crankcase oil or lubricating grease may be used. These adhere fairly well to most surfaces and can be covered with earth or similar loose materials. Remove from metal, glass, and similar smooth surfaces with organic solvents.

(3) Emergency adhesives may be prepared by boiling some types of seaweed in water. The type producing the best results is the reddish, or reddish-green, branchy seaweeds. Green, unbranched seaweeds are devoid of mucilaginous constituents. Brown seaweeds, such as the kelps, when boiled in an alkaline solution, such as leaching potash or soda from wood ashes, yield a good adhesive. All types are capable of temporary binding power. They tend to lose their adhesive quality during rain and fog. Most of these adhesives are best applied in a hot state because they gel quickly at body temperature. Dried and fresh seaweed are equally satisfactory as a raw material. Complete extraction of the adhesive takes place in from 15 to 30 minutes' boiling time.

(4) Native resins and gums are also sources of emergency adhesives. The identity of a gum or a resin-productive species may be determined in any locality from native practices, by consulting proper local governmental agencies, or, as a last resort, by practical experimentation, usually by cutting the bark of trees and shrubs and observing the results. Usually gums and

Table XXX.—Field-expedient green vegetation adhesive.	
Availability	Source: Field-expedient adhesive obtained in small quantity from practically any green vegetation.
Preparation	BILL OF MATERIALS: One or more 5- or 10-gallon metal buckets. One G.I. can or 55-gallon drum. PROCEDURE: Use only freshly cut plant materials; otherwise the quality of the adhesives will be unsatisfactory. Twelve to fifteen 5-gallon bucketsful of vegetation yield about 1 gallon of dark-colored adhesive. Remove larger stalks and woody branches from the vegetation. Pack firmly into 5- or 10-gallon metal buckets until two-thirds filled. Cover material with water to a depth of 2 or 3 inches. Boil over hot fire for about half an hour until greater part of adhesive substances is extracted and a thin syrup is formed. Pour this thin syrup into carrying bucket; discard spent material; repeat to this point as required. Pour contents of carrying bucket into a larger container, such as a G.I. can or empty gasoline drum (fig. 4), and boil down to desired concentration. To avoid boiling over, this container is never more than half filled.
Application	Surfaces on which applied: Metal (as vehicles). Glass. Wood. Fabrics. NOTE: All surfaces must be free from oil and grease. This material is not capable of large-scale production and is suitable for use only on small areas.
How applied	With rags. With improvised brushes or dried grass stalks or pine needles.
Coverage	1,000 square feet per gallon on smooth, non-porous surfaces. 400 to 500 square feet per gallon on rough surfaces.
Climatic limitations	Heat and cold have practically no effect on this adhesive. Fairly good weather resistance except under— Prolonged exposure to rain. High relative humidity.
Durability	Good bonding strength.
Types of granules suggested	With adhesive of the consistency of molasses: Sand. Ashes. Soil. Woods litter. With adhesive of the consistency of a paste, so heavy it will not run after it is applied: Leaves. Twigs. Grass. Other heavier materials. Bark.
Application of granules over this adhesive	Apply immediately. Dust on a second application before the adhesive dries. Apply heavy paste adhesive in small patches, the texturing material is pressed on each patch before the next one is applied.

TABLE XXXI.—Field-expedient starch adhesive.	
Availability	Source: Wheat flour, ground grains. Whole grain, seeds, other hard grai Potatoes, non-fibrous vegetables, fruits. Any other foods that disclose presence of starch.
Preparation	When using flour or ground grains: Stir one part of flour into 4 parts of water by volume until no lumps remain. Heat and continue stirring until mixture becomes thick and smooth. Stir in water while cooling to reduce paste to right consistency. When cooking is out of question, stir flour into cold water. When using whole grain, seeds, other hard sources: Add 4 to 5 parts of water by volume for each part of starchy materials. Boil until material softens and breaks up easily. Add water to replace evaporation. Drain off excess water, mash material, and stir until a smooth paste is formed. Hulls, bark, etc., may be removed by straining through burlap or coarse material. Stir in water to reduce to right consistency. When using potatoes and non-fibrous vegetables: Remove skins or outer coverings; place inner starchy parts in a cooking utensil or container. Add enough water to cover and cook until softened. Pour off excess water; mash remaining material thoroughly and stir to a smooth paste. Stir in more water to adjust paste to right consistency. If skins are not first removed, they can be strained out after mashing. When using fibrous materials, such as cassava, arrowroot, and sago palm: Grate material in improvised grater made by punching nail holes in a bucket cover. Mix grated material with twice its volume of water. Squeeze through a strong cloth. Repeat mixing and squeezing 3 times. Allow strained mixture to stand 2 or 3 hours until starch grains have settled to bottom. Adjust amount of water until it is about 6 times the volume of starch and cook over fire, stirring continuously until smooth, lump-free paste is formed. Add water if thinner consistency is needed.
Application	Surfaces on which applied: Wood. Metal. Fabric.
How applied	With rags. By hand or paddles. Improvised brushes of dried grass stalks or pine needles.
Coverage	Specific data not available.
Climatic limitations	Heat and cold have practically no effect on this adhesive. Tends to peel off smooth surfaces exposed to dry heat.
Durability	Excellent on rough surfaces. Good on smooth surfaces.
Types of granules suggested	Sand. Light colored soil. Other similar materials.
Application of granules over this adhesive	Apply immediately. A second dusting just before drying is advisable to give more coverage. Where a heavier coating is needed, mix together the sand and the starch adhesive into a thick paste, then spread on with hand or paddle, and finally dust over with dry sand while surface is still moist.

Table XXXII.—Glue-paste adhesive.	
Availability	Source: Field expedient for use in continental United States.
Preparation	PROCEDURE: Add enough flour paste to a solution of glue sizing to thicken it. Then emulsify ¾ gallon of rosin sizing varnish in a solution of ½ pound of borax and 2 gallons of hot water. The glue-paste solution is then mixed with the emulsion in equal proportions by volume to form the adhesive.
Application	Surface on which applied: This adhesive is recommended as an emergency substitute for bituminous-emulsion adhesive when applying texturing materials to chicken wire.

resins are plant exudations, caused by injury to the bark. Some are soluble in water, others only in alcohol. To gather gums or resins, strips of bark are usually cut to start the flow of the gummy sap. The flow is usually slow; from 2 days to 3 weeks must pass before it can be collected by chipping it off with a hatchet. In trees with large trunks, pockets to collect the sap can be notched in the trunk just under the places where the bark has been cut. This sap can be dipped out and immediately used without being boiled down. In many cases a fair quantity of sap can be obtained in a few hours.

(5) In some areas, fossiliferous resins, which are found in the ground in depths ranging from 2 inches to 3 feet, are a plentiful source of emergency adhesive.

(6) Natural bitumen is another source of adhesive found on the ground surface in certain areas of Portugese East Africa, in South Australia, and in the Aleutian Islands.

c. All field-expedient adhesives are emergency substitutes; these are not as satisfactory as the standard class-A bituminous emulsion, asphalt cement, or cutback asphalt. Before it has been decided to use these substitutes on a large-scale installation, consideration must be given to the amount of time material, and manpower needed to improvise them.

SECTION III. TEXTURING MATERIALS

10. GENERAL.—Texturing materials are spread over a surface, either loosely or made to adhere. Their most common use is to provide texture for smooth paved surfaces (figs. 5 ① and ②). For this purpose, materials are needed which are durable under abrasion. Texturing materials for non-traffic-bearing surfaces such as roofs, may have less durability. Materials available locally should be used, care being taken to select the one best suited to the purpose at hand. Because a great many materials may be used for texturing, it is impractical to list all. Table XXXIII and figures 6 and 7 give information about a sufficient variety to permit the selection of a material which has characteristics similar to one of the materials described in the

FIGURE 5 ①.—Two-dimensional textured pattern on air-base runway is rolled to force granules into adhesive coating.

FIGURE 5 ②.—A textured section of roadway, showing that a road may be made to blend with the natural texture of the area.

Figure 6 ①.—An oblique view of a surface textured with rock chips.

table and which may be available in quantity locally. They are listed in approximate order of their desirability for traffic-bearing surfaces.

a. To determine the availability of native materials in civilized regions, investigate the byproducts and waste of local factories, mills, shops, slaughter houses, and mines. In uncivilized regions, most likely sources of materials are salvage depots, quartermaster depots, and local agricultural enterprises. No texturing materials are items of issue.

b. When a concrete surface is laid texture can be built in by brushing the surface to expose aggregate just prior to final set. Both concrete and bituminous surfaces can be textured by applying a bituminous seal coat covered with selected aggregate. Textures and colorings may be varied by the selection of aggregate possessing different colors and various degrees of coarseness. This type of surface has good texture, but it also has an abrasive effect.

11. PREPARATION.—*a. General.*—Most materials are applied in the state in which they are purchased. However, the Corps of Engineers has developed a mobile wood-chipping machine which is capable of preparing chips in desirable sizes.

b. Coloring.—When materials do not match the color of the surround-

FIGURE 6 ②.—An oblique view of a surface textured with burlap scraps.

ings, they may be colored before application or painted after application. Coloring before application may be done by mixing with paint in a concrete mixer.

12. APPLICATION.—When asphalt cement is used as the adhesive, the texturing material selected is applied immediately following the application of the adhesive coating. When the adhesive is bituminous emulsion or a cutback, the texturing material is applied shortly after the adhesive has become tacky—normally 10 to 15 minutes later. Texturing materials are applied in various ways. The most expeditious method is to use a mobile wood-chip distributor, as described in paragraph 56. Another method is to dump piles at the sides of the area to be covered and distribute them by hand shovel. Another method is to spread them from the back end of dump trucks. The trucks are driven in reverse so as not to run over the adhesive surface itself. As soon as the materials have been spread, they should be rolled with a heavy rubber-tired roller or a 5-ton, or heavier, road roller (fig. 5 ②). This is essential in order to force the granules deeply into the adhesive. The surplus granules may be removed by sweeping after the adhesive has set.

TABLE XXXIII.—Texturing materials.

Texture Material	Durability Compared With Rock Chips	Abrasion	Average Light Meter Readings, Vertical, on Bright Sunny Days	Coverage per Ton in Square Yards	Sources	Remarks Based on: 1. Abrasion 2. Durability 3. Texture
Wood chips, 1″	75%	Low	25	800	Tanneries, wood pulp concerns, paper mills.	Very good.
Rock chips, ½″	100%	High	20	90	Quarries and road building contractors.	Very good.
Gravel	100%	Medium	40	90	Gravel pits.	Fair.
Leather chips	75%	Low	30	115	Shoe and leather goods manufacturers.	Very good.
Coffee beans	75%	Low	Not available	90	Coffee growers.	Very good.
Wood cubes	60%	Low	25	161	Lumber (cubes made on site)	Fair.
Fabric (Burlap Scraps or Similar Materials)	30%	Low	30	2,000	Rag and textile concerns.	Good.
Plant mix, 1¼″ Open Graded Aggregate	100%	Medium	35	22	Asphalt distributors.	Very good.
Slag	80%	High	37	90	Iron and steel blast furnaces.	Very good.
Pine bough tips	20%	Low	25	400	Local labor.	Fair.
Pine cones	20%	Low	Not available	161	Local labor.	Good.
Excelsior	30%	Low	25	2,000	Wood pulp concerns.	Very good.
Shredded coconut fiber (Coir).	15%	Low	Not available	2,000	Byproduct of cocoa manufacturing.	Good.

Redwood fiber	30 %	Low	30	1,600	Redwood Fiber Co., Santa Cruz, Calif.	Good.
Cinders	90 %	High	Not available	90	R.R. round houses, boilers, factories.	Fair texture; good durability.
Feathers	15 %	Low	40	20,000	Poultry slaughter houses, mattress manufacturers.	Fair.
Cotton seed hulls	50 %	Low	Not available	3,000	Cotton gi	Good texture; poor durability.
Spanish moss	10 %	Low	Not available	2,000	Local labor.	Excellent texture; poor durability.
Paper, shredded	05 %	Low	Not available	2,000	Paper mills, junk dealers.	Good texture; poor durability.
Sawdust, ⅛″	75 %	Low	Not available	3,000	Lumber mills, furniture manufacturers, plywood mills.	Poor texture.
Leaves and chopped palm fronds	15 %	Low	Not available	2,000	Local labor.	Fair texture; poor durability.
Seaweed	10 %	Low	Not available	2,000	Local labor.	Fair.
Tin cans			Not available		Salvage depots.	Very good texture.

NOTE: These figures are approximate, for purpose of comparison and rough estimates.

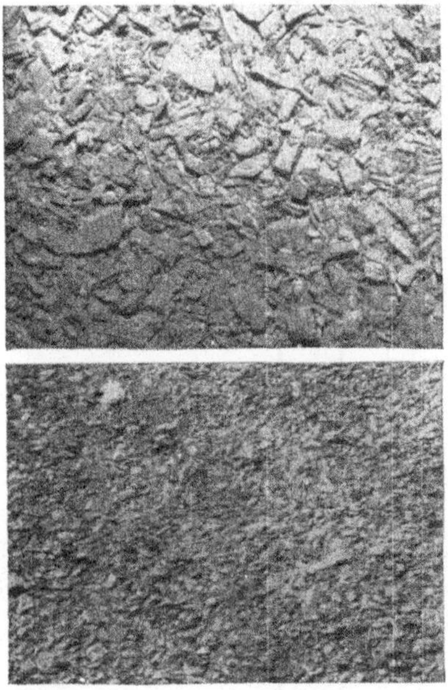

FIGURE 7 ①.—Wood-chip texturing; vertical and oblique views.

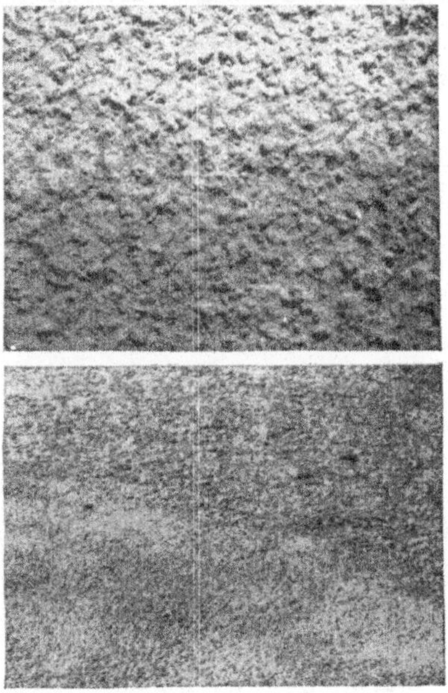

FIGURE 7 ②.—Slag, vertical and oblique views.

APPLIED SURFACE

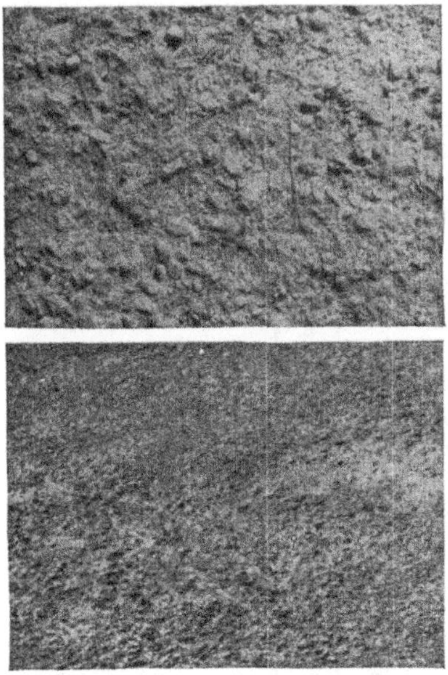

FIGURE 7 ③.—Gravel, vertical and oblique views.

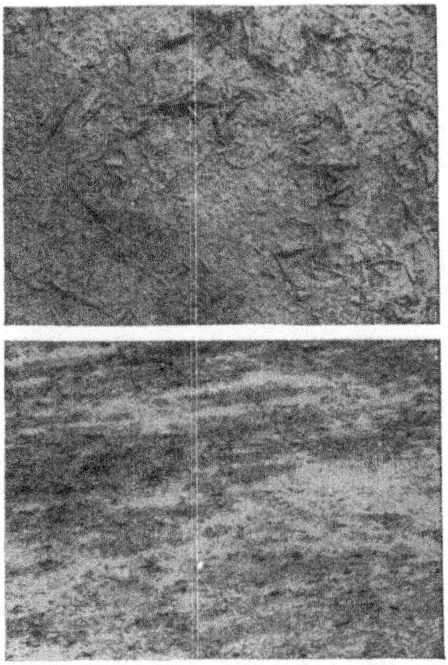

FIGURE 7 ④.—Redwood fiber, vertical and oblique views.

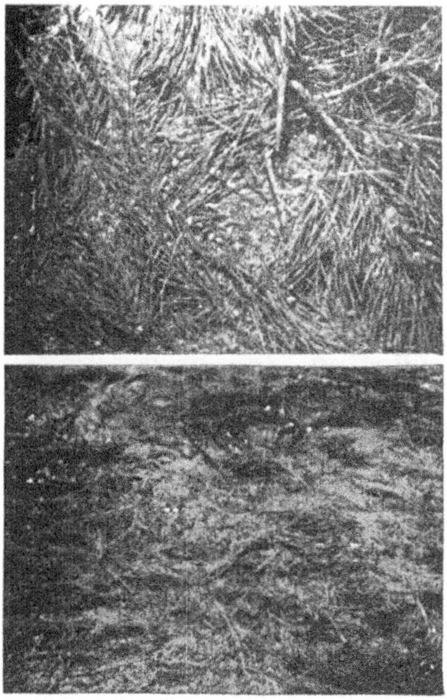

FIGURE 7 ⑤.—Pine-bough-tip texturing, vertical and oblique views.

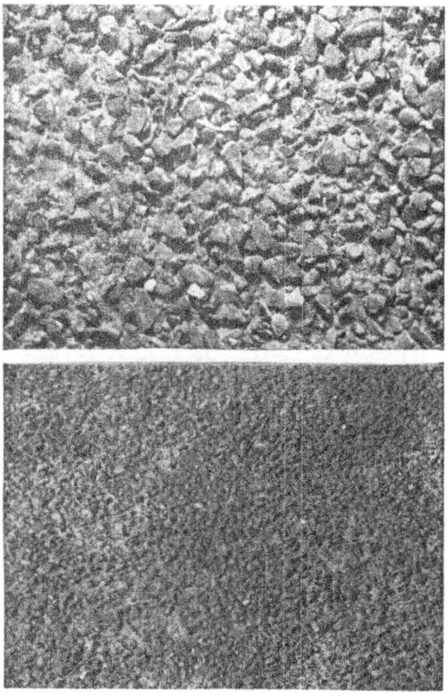

FIGURE 7 ⑥.—Rock chips, vertical and oblique views.

TEXTURING MATERIALS

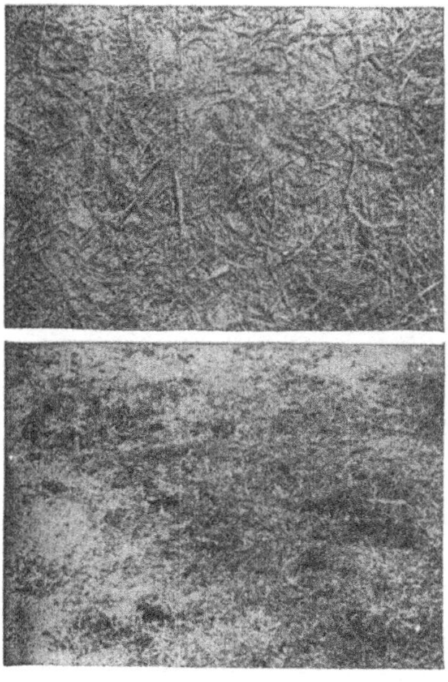

FIGURE 7 ⑦.—Excelsior, vertical and oblique views.

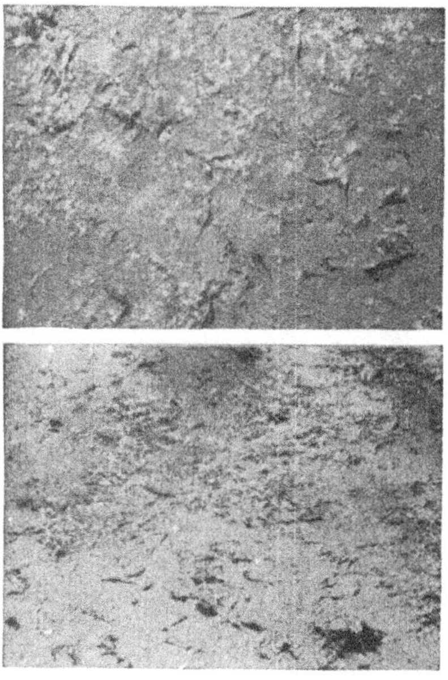

FIGURE 7 ⑧.—Feathers, vertical and oblique views.

SECTION IV. SCREENING MATERIALS

13. GENERAL.—Screening materials are used for blending or hiding or for deception. They are most frequently employed to blend with a natural background, as extending natural overhead or lateral cover, and to add to or to construct a man-made object, as adding a false wing to an existing building or making decoy vehicles.

14. CLOTH AND CLOTH SUBSTITUTES.—*a. General.*—Screening materials include issue cloth, shrimp nets, commercial substitutes, and field expedients.

b. Cloth is a common type of material available for screening (fig. 8 and 9). It is especially useful in decoy construction. Issue impregnated camouflage fabrics have infrared reflectance and are mildew-proofed.

c. Substitutes for cloth may be found in various locations; for example—

(1) Chen-chen, a durable wood-fiber cloth in sheets up to 6 by 8 feet may be obtained in West Africa through the forestry department.

(2) Palm matting can be obtained in some areas in ready-made mats, or it can be ordered in any specified size. It is light weight and can be supported on twine netting.

(3) In the Far East straw matting may prove useful in decoy construction.

(4) Birch bark has been used for decoy construction in Alaska.

(5) Roofing paper and cardboard are limited substitutes for cloth.

(6) See paragraph 17*c* for small-mesh wire netting, which may be used as a surfacing material when painted.

15. SMALL-MESH NETS.—*a. General.*—These nets require no garnishing to provide concealment. They form a veil which masks equipment hidden under them, breaking up its form, dulling its brightness, and throwing irregular shadows over it. As a drape, the small-mesh net affords effective concealment of identity from visual observation. These nets cannot be used on flat-tops since they do not thin out at edges.

FIGURE 8.—Burlap cloth, actual size.

Table XXXIV.—Cloth.	
Burlap	Osnaburg
Issued in bales containing 2,000 yards. Issue bale commonly contains bolts 40 inches wide and approximately 100 yards long. Issue colors and stock numbers: Olive drab 49-2405.4-65 Natural tan 27-2590.075-4 Specifications: Federal specification CCC-B-811 (Untreated); Corps of Engineers Tentative Specification No. T-1212 (Impregnated) NOTE: This specification may shortly be replaced by EBP No. 396, which adds flame resistance to the other properties called for in T-1212.	Issued in bales containing 1,000 yards. Issue bale commonly contains bolts 40 inches wide and of variable lengths. Issue color and stock number: Natural 27-3014.5-4 Specifications: U.S. Army Specification 6-280 (Untreated); Corps of Engineers Tentative Specification No. T-1212 (Impregnated). NOTE: This specification may shortly be replaced by EBP No. 396, which adds flame resistance to the other properties called for in T-1212. This is a substitute issue for burlap; it is not being purchased at present; some quantities may still be in stock.

Other sources for cloth are:

Local weavers or textile manufacturers.

Usable camouflage cloth may be found in military depots, salvage depots, and abandoned civilian homes in theaters of operations. Such items as draperies, bedspreads, and table cloths are particularly useful. Select materials that resemble burlap. Colors can be changed by re-coloring with issue camouflage paints.

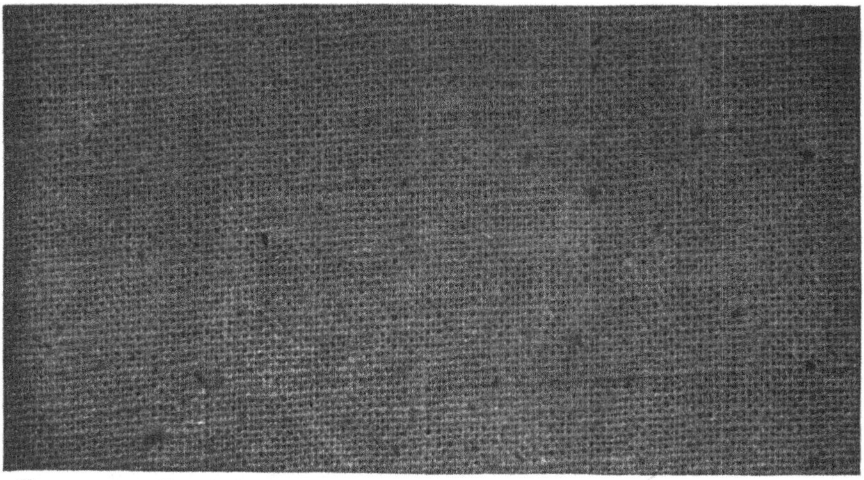

Figure 9.—Osnaburg, a cotton cloth, actual size.

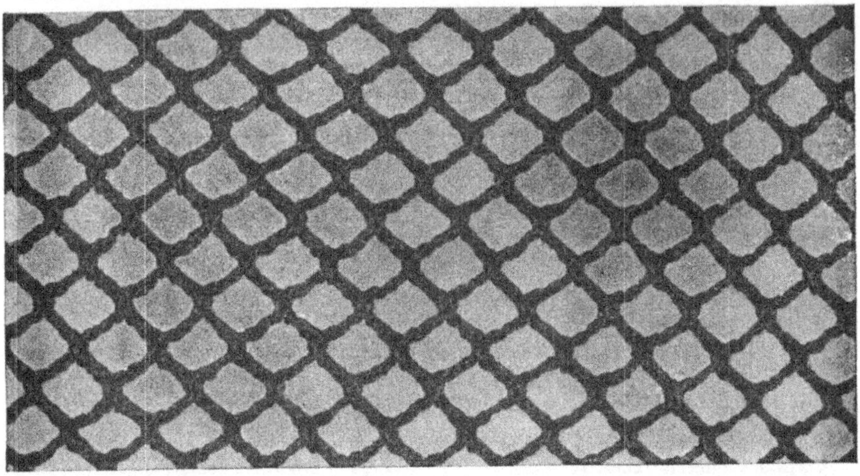

FIGURE 10.—One type of shrimp net, actual size.

b. Issue.—The issue small-mesh net is shrimp net, which is a closely woven cotton fabric with meshes of from ¼ to $^5/_{16}$ inch square (fig. 10).

c. Substitutes for shrimp nets.—(1) Nets of scrim (a material similar to mosquito netting in weave and appearance) may be pattern-painted on both sides to match the coloring of different sites (fig. 11). Sources of scrim are theatrical supply houses, textile manufacturers, and retail merchants. Common width for scrim is 39 inches; the length is variable. It is ordinarily light yellow in color.

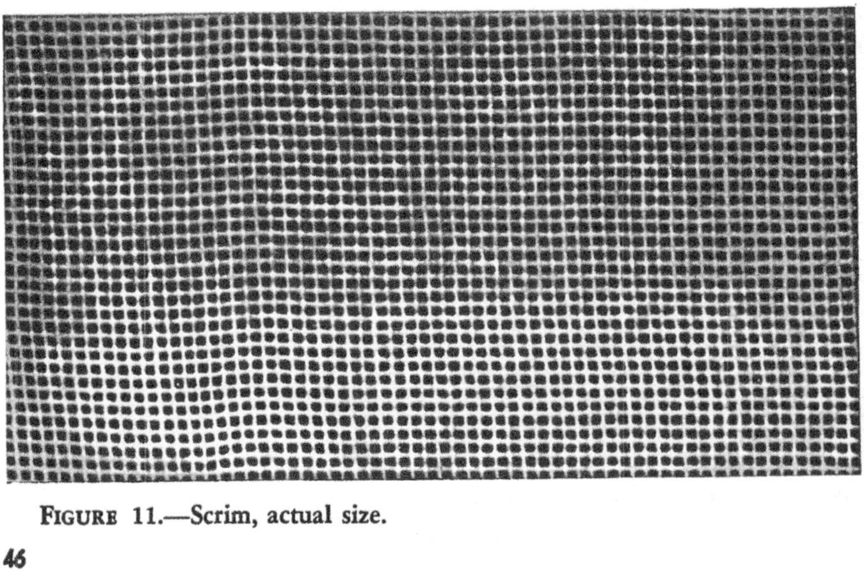

FIGURE 11.—Scrim, actual size.

Table XXXV.—Small-mesh nets.				
Size (feet)	Weight (pounds)	Cubage (cubic feet)	Color	Stock Number
22 by 22	30	2.0	Olive drab Sand	49-4612.45-65 49-4612.45-75
29 by 29	55	3.5	Olive drab Sand	49-4612.5-65 49-4612.5-75
36 by 44	100	5.5	Olive drab Sand	49-4612.6-65 49-4612.6-75
45 by 45	125	6.0	Olive drab Sand	49-4612.75-65 49-4612.75-75
Specification: Corps of Engineers Tentative Specification No. T-1669.				

(2) Mosquito netting, which is similar to scrim, may be made up into nets as a substitute for shrimp nets. It is not issued, but may sometimes be obtained from commercial sources.

(3) Woven-paper netting is a material similar in appearance to shrimp netting. It is made of tightly wound strands of impregnated, exceptionally strong paper, woven in the same manner as shrimp nets. Sources of supply are commercial manufacturers of onion-sacking materials and the like.

(4) In the absence of shrimp nets, commercial manufacturers of lace fabrics or curtain materials can be engaged—in some theaters of operations—to weave similar nets. By specifying *Raschel, Terchen, Lovers Lace, or Nottingham lace-type fabrics,* of the same size as shown in figure 10, weavers will be able to produce a satisfactory substitute by converting their knitting or lace-making machinery.

Figure 12.—A section of a field-expedient mat screen, erected as a vertical ground screen. Note arrangement of guy wires.

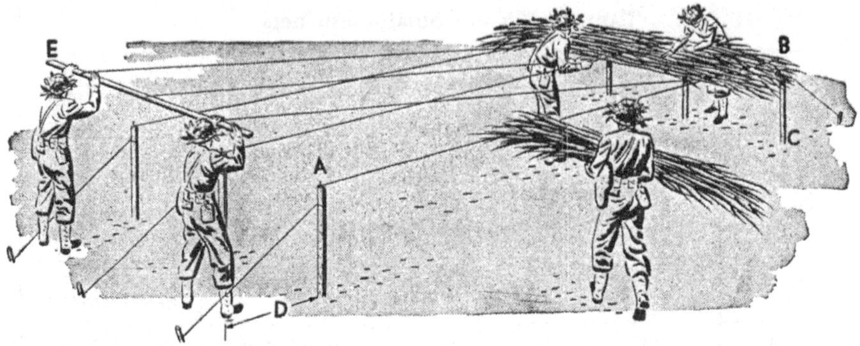

FIGURE 13.—Improvised mat loom, showing method of operation.

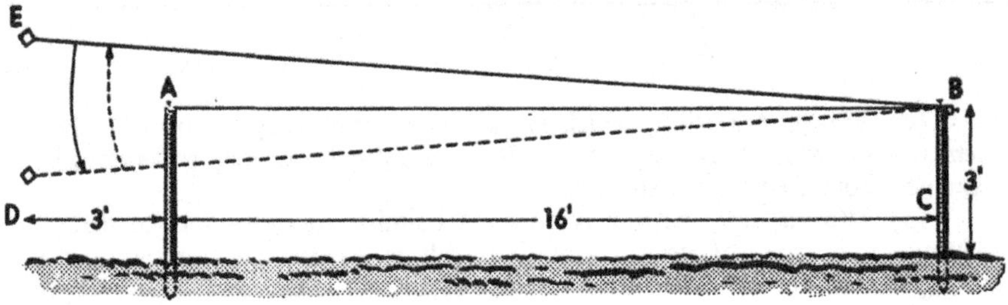

FIGURE 14.—Method of weaving a mat screen.

16. FIELD-EXPEDIENT MAT SCREENS (fig. 12) for making imitation native thatched houses, imitation thatched roofing, and false hedges are woven on an improvised mat loom of local materials, such as brush, corn stalks, palm fronds, sugar cane, reeds, or thatch.

a. Improvised mat loom.—Posts are set firmly in ground at 2-foot intervals and guyed to stakes, as in figure 13. Wider mats require additional pairs of posts to accommodate extra wires.

b. Method of weaving (fig. 14).—Fasten strands of 10-gage wire to nails in top of each of three posts, A, stretch to a sapling slightly longer than the width of the desired screen, B; pass each wire around sapling, which is then nailed to top of posts, C. Leave nail heads exposed for easy removal. Continue wires back to D, a point 3 feet beyond posts, A. The wires are then fastened to another sapling, E. Two men alternately raise and lower sapling, E, as reeds or other materials are inserted in about 6-inch sections. Rack wires between alternate sections. When mat screen is completed, cut wires at A and E and fasten sapling, E, to edge of screen by the loose ends of the cut wires.

17. OTHER SCREENING MATERIALS.—*a. Truck tarpaulins* are occasionally useful where an opaque and waterproof fabric is needed. For example, they have been used over a supporting framework to cover dug-in positions, as in figure 15. Camouflage over the tarpaulin must be added.

FIGURE 15.—An improvised shelter for a vehicle, constructed by digging in and erecting a supporting overhead frame, which is covered with a tarpaulin and camouflaged with natural materials.

FIGURE 16.—Snow fencing erected to blend a railroad gun with a sand-dune background. An excellent matching of color and texture results from dusting local sand over freshly painted snow fencing.

b. Snow fencing (fig. 16) is a means of obtaining overhead screening. As used by highway departments, this flexible fencing is erected vertically; for camouflage, normally, the fencing is supported horizontally or at an angle to the ground. A semi-permanent installation is shown in figure 16. It is effective only against high-altitude observation. Snow fencing is obtainable commercially from manufacturers of fencing in rolls 4 feet wide and 100 feet long, weighing 250 to 300 pounds. It is made of wooden slats, each ½ inch by 1½ inches by 4 feet, which are interwoven—small gaps being left between slats—by 5 double strands of 12½-gage wire. The fencing comes stained red or unstained, and must be painted or textured, as by the application of sand to the wet paint. Because of its weight, adequate support must be provided.

c. Fine-mesh screening is placed over observation and firing apertures in various camouflaged positions, such as those simulating rocks, to form removable screens which break up the shadow of the opening and do not obstruct vision. Salvaged wire fly screening, dyed mosquito netting, or similar closely-woven mesh net is used for this purpose. Galvanized 16-gage wire screening may also be used as a surfacing material in decoy construction and when painted white is excellent in concealing small installations in snow conditions.

SECTION V. SUPPORTING MATERIAL

18. LARGE-MESH NETTING.—Large-mesh netting requires the addition of garnishing material to fill in the openings; small-mesh nets (par. 15) do not. Large-mesh netting, made of twine or wire, is available either ungarnished or pregarnished. Whenever garnishing does not blend with the terrain where it is to be used, garnishing must be spray-painted or replaced.

a. Twine nets.—(1) Twine nets serve either as camouflage drapes or

FIGURE 17.—A captured Italian net, garnished with tufts of grass, erected to provide overhead concealment.

TABLE XXXVI.—Netting, Twine.

NETS, TWINE, UNGARNISHED.

Mesh: 2¼ inches square.

Size (feet)	Weight (pounds)	Cubage (cubic feet)	Stock Number
15 x 15	4.0	0.3	49-5644.5-3
14 x 29	6.7	0.5	49-5644.5-25
17 x 35	10.0	0.9	49-5644.5-35
29 x 29	12.2	1.1	49-5644.5-5
36 x 44	24.0	1.8	49-5644.5-65

Corps of Engineers Tentative Specification No. T-1534.

These obsolete sizes may still be available in limited quantities at some engineer depots: 12 x 12 feet, 22 x 22 feet, 30 x 30 feet, 45 x 45 feet.

NETS, TWINE, FABRIC GARNISHED (*See* fig. 18).

Issue.

Mesh: 2¼ inhes square.

Size (feet)	Weight (pounds)	Cubage (cubic feet)	Stock Number
15 x 15	9.0	0.9	49-4956.3
14 x 29	12.0	1.0	49-4956.25
17 x 35	20.0	2.0	49-4956.35
29 x 29	27.0	3.3	49-4956.5
36 x 44	51.0	5.0	49-4956.65

Colors: Add "—3" to stock number to obtain desert garnishing.
Add "—5" to stock number to obtain summer garnishing.
Add "—7" to stock number to obtain winter garnishing.

Corps of Engineers Tentative Specification Number:
T-1534 (Twine Netting)
T-1220 (Garnishing)

These obsolete sizes may still be available in some engineer depots: 12 x 12 feet, 22 x 22 feet, 30 x 30 feet, 45 x 45 feet.

covers for flat-tops in temporary installations. For comparative data on twine nets, see Table XXXVI. Garnishing materials and manufacturing methods are described in chapter 2, section VI. It is characteristic of twine nets that they shrink when wet and expand when they dry out. Nets issued prior to 1942 employ a diagonally woven mesh; they have a tendency to

FIGURE 18.—Twine netting garnished with burlap, two-thirds actual size.

pucker and become irregular in shape. These nets require more intermediate support than do the present issue nets, which have a squarely woven mesh. Alteration of diagonally woven nets is described in paragraph 19*f*.

(2) Special nets belonging to net sets are covered in chapter 5, section II.

(3) Nets are organic equipment for many units; if destroyed or lost, suitable substitutes may often be found in captured enemy nets (fig. 17), salvaged Allied nets, or fishing nets purchased or requisitioned from local fishermen, seiners, or manufacturers. In some areas, local skilled labor may be employed to manufacture nets from native materials.

b. Wire netting does not sag or deteriorate with use. It is suitable for semi-permanent installations. Wire netting is available either ungarnished or pregarnished with steel wool, glass fiber, or fabric (table XXXVII). It is rust-resistant and non-inflammable, which makes it especially desirable

FIGURE 19.—Wire netting garnished with steel wool, two-thirds actual size.

FIGURE 20.—Manufactured, feather-garnished wire netting, two-thirds actual size.

FIGURE 21 ①.—Wire netting garnished with glass fiber to resemble the texture of grass, two-thirds actual size.

②.—Wire netting garnished with glass fiber to resemble the texture of foliage, two-thirds actual size.

TABLE XXXVII.—Netting, Wire.

NETTING, WIRE, ENAMELED OR GALVANIZED, STEEL-WOOL GARNISHED (*See* fig. 19).

Size (feet)	Weight (pounds)	Cubage (cubic feet)	Color	Stock Number
6- by 75-foot rolls	100	6.4	Dark green	49-6676.02-3
			Earth brown	49-6676.02-35
			Field drab	49-6676.02-5
			Light green	49-6676.02-55
			Olive drab	49-6676.02-65
			Sand	49-6676.02-75

Corps of Engineers Tentative Specification No. T-1284.

NOTE: This material is bonderized to prevent rusting, but in humid climates this is not sufficient to prevent rusting, and the wire and steel wool have a short useful life.

NETTING, WIRE, STEEL, GARNISHED WITH CHICKEN OR TURKEY FEATHERS (*See* fig. 20).

Size	Color	Stock Number
6- by 150-foot rolls	Dark green	49-7150.05-3
	Earth brown	49-7150.05-35
	Field drab	49-7150.05-5
	Light green	49-7150.05-55
	Olive drab	49-7150.05-65
	Sand	49-7150.05-75

Specification: Engineer Board Procurement No. 409.

For field manufacturing technique, see paragraph 23*d*.

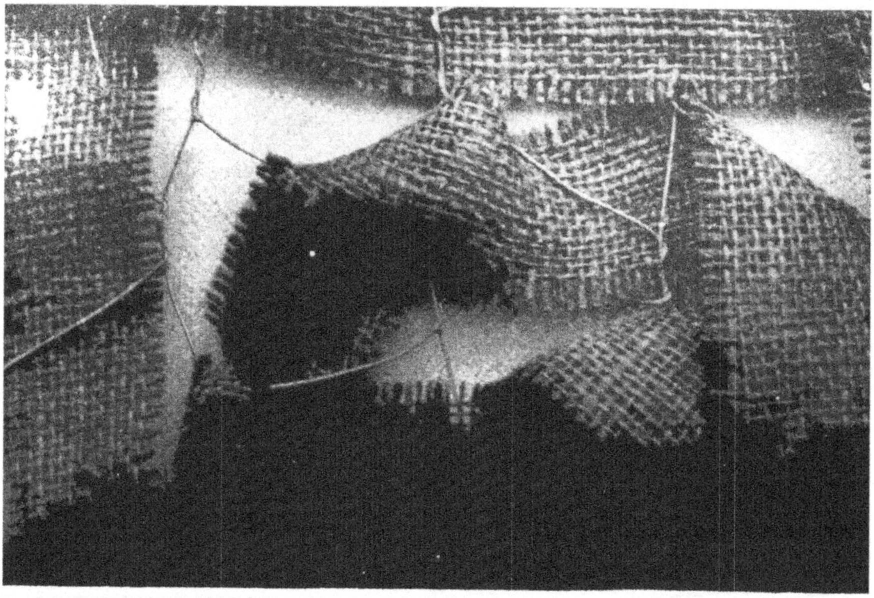

FIGURE 22.—Fabric-garnished wire netting, two-thirds actual size.

TABLE XXXVII.—Netting, Wire. *Continued.*

NETTING, WIRE, GLASS-FIBER GARNISHED (*See* fig. 21).

Issued in two types: with grass texture; with foilage texture.

Size (feet)	Weight (pounds)	Cubage (cubic feet)	Color	Grass Texture	Foliage Texture
6- by 75-foot rolls	about 100 pounds	6.4	Dark green	49-7100.1-3	49-7103.1-3
			Earth brown	49-7100.1-35	49-7103.1-35
			Field drab	49-7100.1-5	49-7103.1-5
			Light green	49-7100.1-55	49-7103.1-55
			Olive drab	49-7100.1-65	49-7103.1-65
			Sand	49-7100.1-75	49-7103.1-75

Corps of Engineers Tentative Specification No. T-1719.

NOTE: This garnishing is superior to steel wool as it is more durable, lighter in weight, fireproof, and does not use strategic materials. The glass wool is enclosed between two layers of chicken wire, the upper being a light 6-inch hexagonal mesh; the weight of the finished article is approximately the same as for rolls of steel-wool-garnished wire netting.

NETTING, WIRE, STEEL ENAMELED OR GALVANIZED, FABRIC GARNISHED (*See* fig. 22).

Size (feet)	Weight (pounds)	Cubage (cubic feet)	Color	Stock Number
6- by 150-foot rolls	95	5.5	Dark green	49-6332.05-3
			Earth brown	49-6332.05-35
			Field drab	49-6332.05-5
			Light green	49-6332.05-55
			Olive drab	49-6332.05-65
			Sand	49-6332.05-75

Corps of Engineers Tentative Specification No. T-1219.

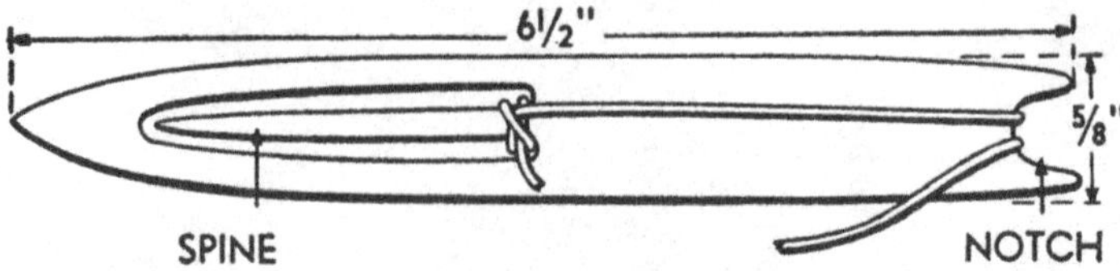

Figure 23 ①.—Netting needle, showing method of attachment of twine.

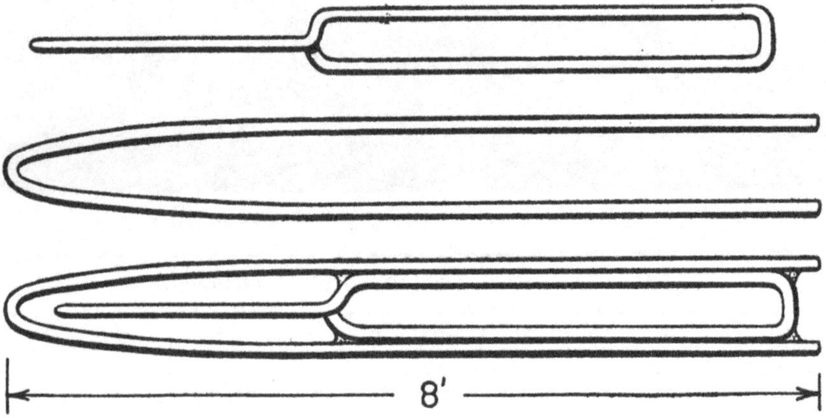

②.—Netting needle, made of No. 10 wire.

for covering inflammable or explosive stores. Because it is both bulky and stiff, it cannot be used successfully as a drape. In mounded or sloped screens, however, wire netting is the best covering material since, once "moulded" to a frame's shape, it will hold that shape indefinitely. Corps of Engineers Tentative Specification No. T-1288 covers welded-wire netting, both zinc- and enamel-coated, and various sizes of mesh which are used for camouflage nets; T-1481 covers camouflage wire netting.

19. REPAIR OF TWINE NETS.—*a.* Torn twine nets must be repaired at the first opportunity. In the field they cannot be quickly replaced and if not mended immediately, further damage will result. Repairs are made by weaving or patching.

b. Materials for repairing twine nets are—

(1) Number 18 (18-thread) medium-laid seine twine. One pound of twine contains approximately 1,000 linear feet. This is a class IV issue.

(2) Pocket knife.

(3) Netting needle, available commercially, or a field-expedient needle made of thin hard wood 6½ to 8 inches long, ⅝ to ⅞ inch thick (fig. 23 ①). Cut out the central spine just long enough so that it must be pushed to one side or the other to permit the loops of twine to pass over its point. This prevents the needle from unwinding when dropped, but permits the user to unwind one loop at a time by a slight pull towards the point of the

needle. A good needle may also be made by soldering together two 15-inch lengths of 10-gage wire after they have been bent to shape (fig. 23 ②). One piece forms the exterior and is bent in a simple elongated "U" shape. The second piece forms the body and spine. It is bent in an elongated, rectangular shape with the spine extending from the center of one end, and the joint is soldered. The long sides of the body are soldered to the exterior wire. Care should be taken that the sides of the U-wire extend beyond the body section.

c. Repairing tears.—(1) The first step in mending a torn or cut net is to trim away the edges of the tear so that weaving is not interrupted by frequent cutting.

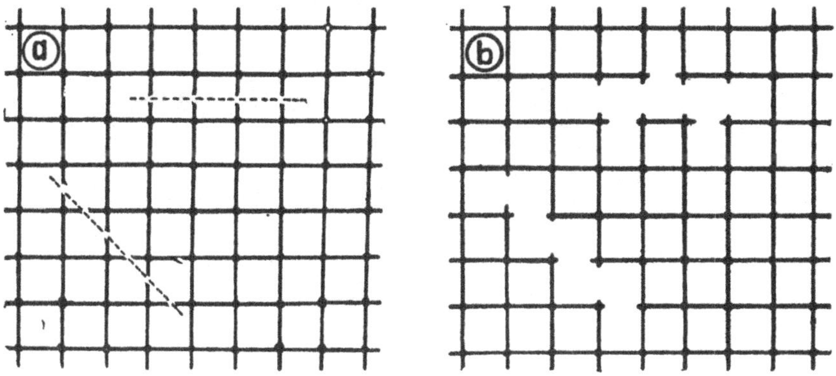

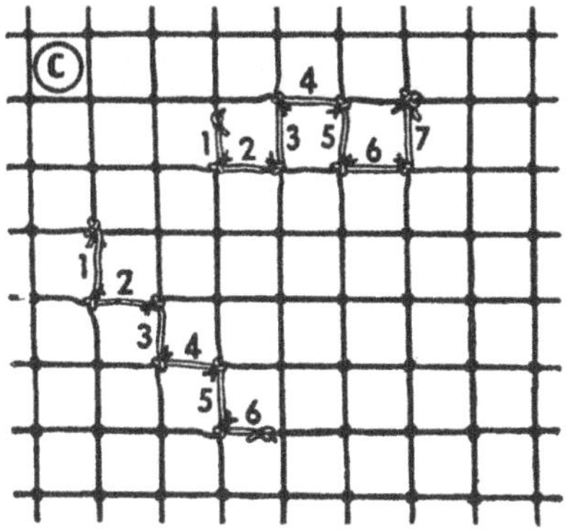

FIGURE 24 (a).—A section of torn net. Dotted lines indicate tears.

(b).—The two tears are properly trimmed.

(c).—The tears are repaired by starting at 1 and then tying at 6 and 7.

(2) The first requirement is that mending must start and finish at a knot joining *three* strands or from a tag end leading from such a knot. This is necessary because only one end of mending twine is attached at a knot, and unless the knot is already composed of three unbroken strands of the original net, the repair will not have the required four strands radiating from each knot.

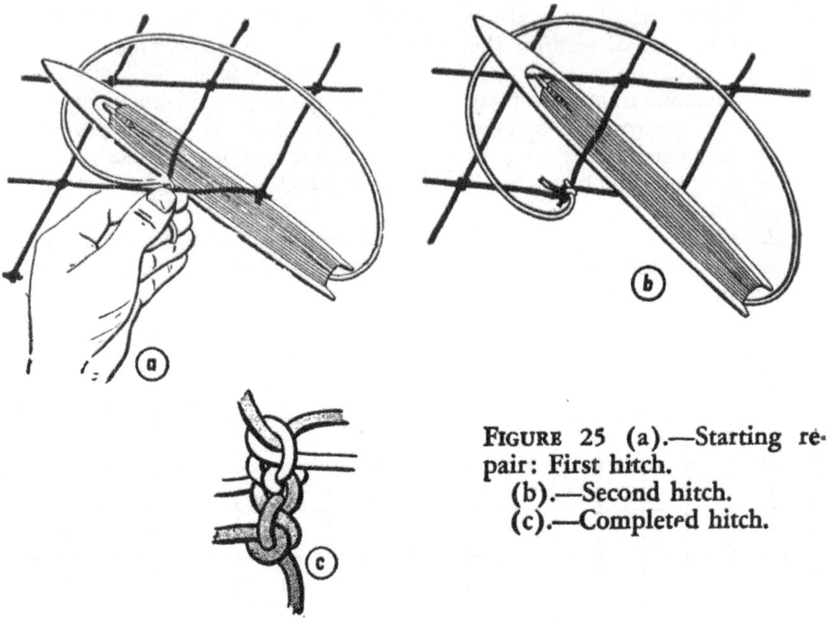

FIGURE 25 (a).—Starting repair: First hitch.
(b).—Second hitch.
(c).—Completed hitch.

(3) The second requirement is that knots around the edges of the tear must have *two* and only two unbroken strands of the original net.

(4) In trimming the tear, unnecessary tag ends are cut off first, then enough other strands are cut out to satisfy the above requirements.

(5) Figure 24 (a), (b), and (c) show different typical tears before trimming, the same tears after trimming, and the sequence in which the tears are woven.

d. Weaving tears.—(1) The end of the twine should be tied on as in figure 25 when the mend starts at a knot where three strands join. Note that the end of the twine is placed between two of the strands, and the second hitch is made around the middle strand only. The second hitch is made this way to bind the end of the twine more securely without excessively distorting the shape of the mesh.

(2) If the mending starts or finishes at a tag end, the twine is tied to the tag end with a square knot.

(3) The sequence of weaving depends upon the shape and position of the tear with respect to the weave of the net and must be determined for each job. Figure 24 (c) shows the sequences for the tears illustrated. The most convenient method for finding the proper sequence of weaving the tear is to spread the net out flat so that the meshes are square; then thread the twine through the meshes—until the proper sequence is found by trial. The twine may then be cut and left in the net to guide the weaving. The guiding twine is removed after the repair is finished. With practice, one will become sufficiently expert to dispense with a guiding twine.

(4) In adjusting a loop, care must be taken to note whether the loop forms one or two sides of a mesh and to adjust the size accordingly.

(5) Some complicated tears cannot be re-woven in a continuous sequence without cutting out an excessive amount of net. In such cases, it is better to trim less extensively and weave several sequences, beginning and ending at three-strand knots, as described above.

e. Patching a large hole.—(1) To mend a net containing a large hole insert a patch cut from a scrap net or weave a piece of net and insert it in the hole. Figure 27 explains how to weave nets.

(2) The first step in patching is to lay the net out flat and pull the meshes square. Then enlarge the hole to a roughly rectangular shape—45° to the mesh—surrounded by knots joining two strands (fig. 26 (a)). Unlike the procedure in mending tears, a three-strand knot is not used for starting or finishing the insertion of a patch. This is because patch weaving starts and finishes at the same knot rather than at different knots as in mending a tear.

(3) A rectangular patch is now cut or woven with one less two-strand knot on each of the sides than is found on the corresponding side of the hole (fig. 26 (b)).

(4) The patch is inserted in the net by weaving continuously around it, as shown in figure 26 (c).

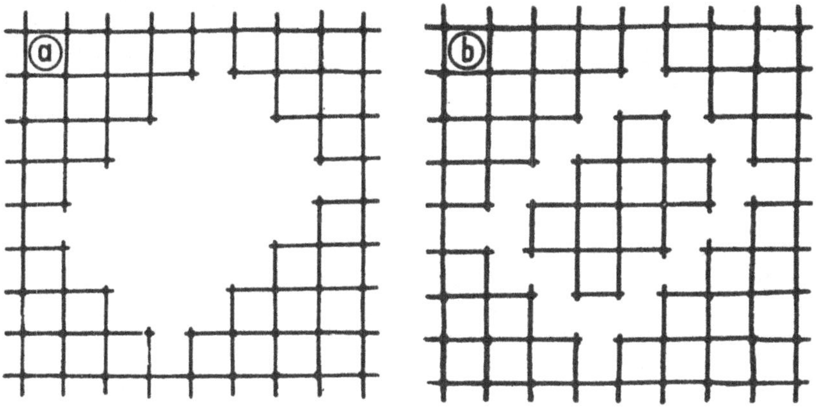

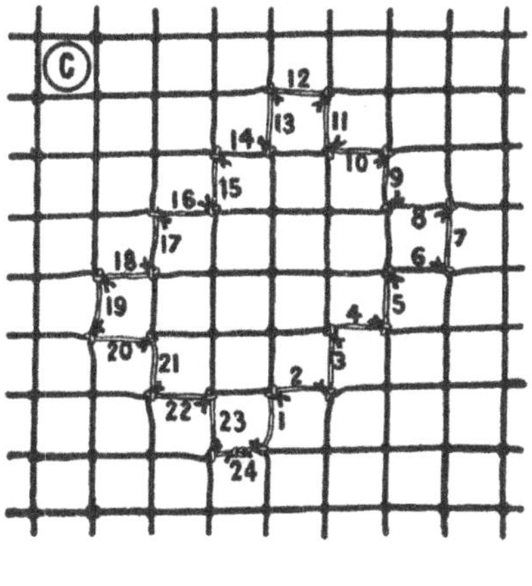

FIGURE 26 (a).—A rectangular hole cut at a 45° angle to the mesh.

(b).—The patch placed in position in preparation for attachment.

(c).—The patch joined to the net by weaving from 1 to 24, where it is tied with a square knot.

f. Alteration of diagonally woven nets.—The rope edging of old-style issue twine nets is often too short and requires lengthening to permit nets to stretch tight when used on flat-tops. To correct this difficulty—

(1) Unsplice and remove rope edging, cutting points of attachment.

(2) Place one man at each end of an edge of net. Pull this edge taut until mesh along it is closed. Measure edge.

(3) Repeat this procedure on each edge of net.

(4) The length of rope needed for each edge is three-quarters the length of the edge.

(5) Re-thread peripheral rope through edge meshes of net and re-splice ends.

HOW TO WEAVE NETS

FIGURE 27 (a).—To fill the needle, tie end of twine around base of central spine with a clove hitch, leading it down one side to notch and up other side to tip of spine. Bend spine enough to permit twine to loop over it and continue to wind in same way. Continue until twine is approximately ¼ inch from end of spine. Leave a free end of about 24 inches of twine.

(b).—Make a loop by tying a bowline in free end of twine. Length of loop when stretched so that sides come together should be same as distance between diagonally opposite knots of a mesh stretched so that the sides come together. This dimension is called the "mesh." A standard camouflage net has a 4½-inch mesh, which means each small square is 2¼ inches on a side.

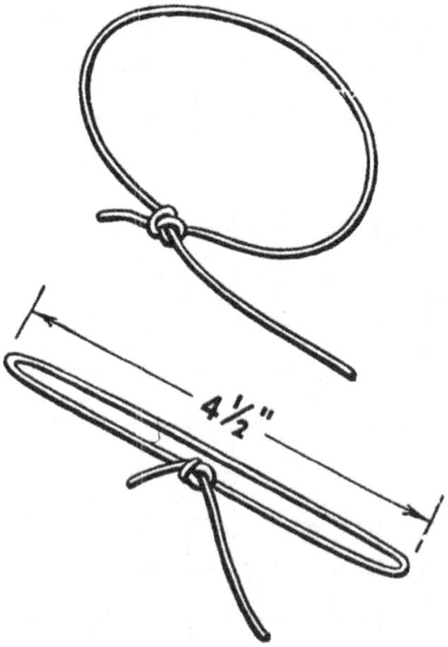

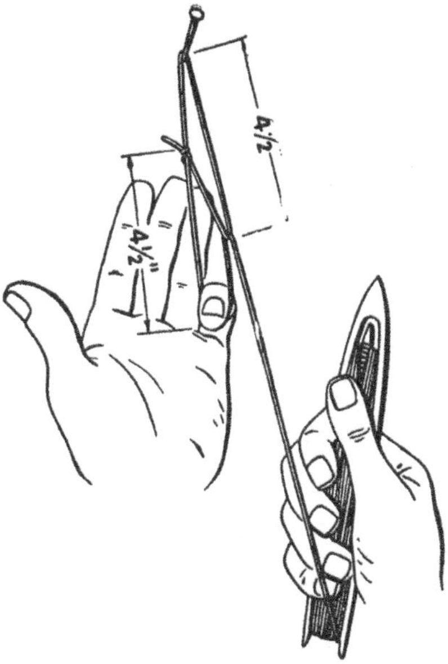

(c).—The next step is to weave a chain of meshes equal to length of net desired. To make a chain, suspend loop just formed from a convenient nail. Turn loop so that knot is in middle of left-hand side. Pass needle up through loop with right hand, regrasp needle with right hand and pull it down towards right hip. Hook left little finger in new loop from behind so loops may be grasped by left hand. Adjust new loop to same size as first one by pulling with needle or with finger. Loops and twine should be stretched firmly toward weaver at this stage.

(d).—When size of loop is adjusted, grasp twine where it passes through first loop with left thumb and forefinger. Note that thumb is *behind* twine leading to needle and that thumbnail grips bottom of old loop. Now throw a loop of twine up to left front with running end leading from top of loop. Pass point of needle *behind* two sides of first loop, *in front* of twine leading down from original bowline to left little finger, and *through* loop just thrown up to left front. Appearance should now be as shown.

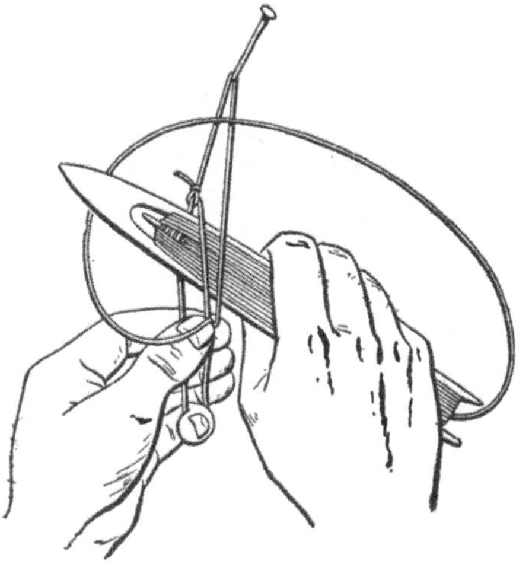

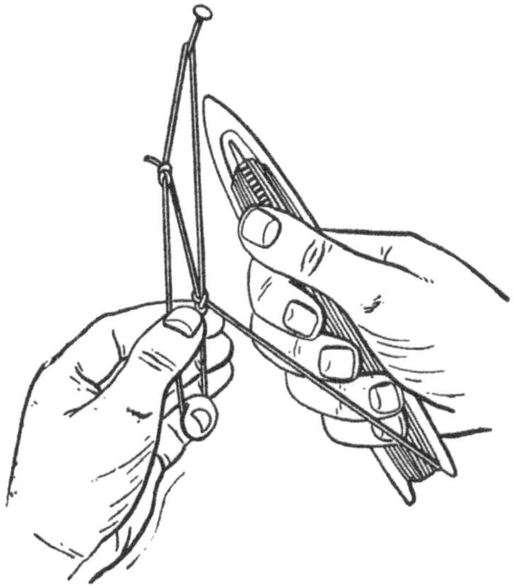

(e).—Regrasp point of needle with right hand and pull knot tight by pulling the needle down towards right hip, at the same time keeping a firm hold with left thumb and forefinger and with left little finger. Mesh is now completed and should appear as shown. Caution: Do not release hold with thumb and forefinger of left hand until knot has been tightened.

(f).—Remove first loop from nail, turn it over and replace it so that twine again leads from middle of left-hand side of loop. Weave another mesh in same manner as first one; turn chain over and continue until number of meshes desired for one side of net have been made. After first few meshes are woven it is no longer necessary to remove end loop from nail and turn the whole chain over. Chain can be twisted until twine leads off from left side. After each mesh is woven, twist chain back to weave next mesh. If net is to be used to patch another net, the exact number of meshes required should be counted. Otherwise, it is sufficiently close to allow 17 to 17½ inches of chain per foot length of edge of net. This is the standard used in making issue camouflage nets and should give a net which stretches tightly when pulled out.

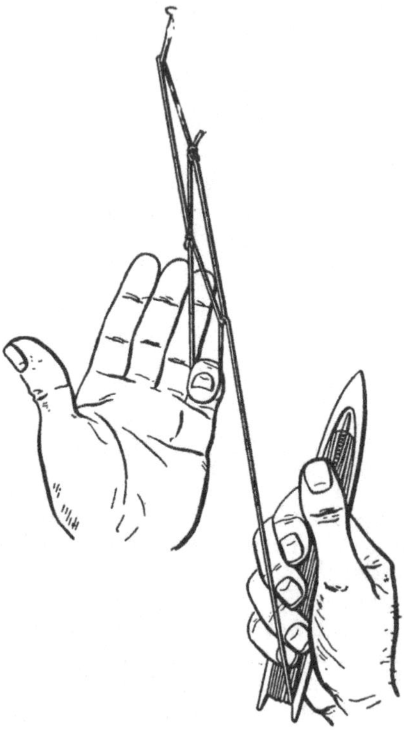

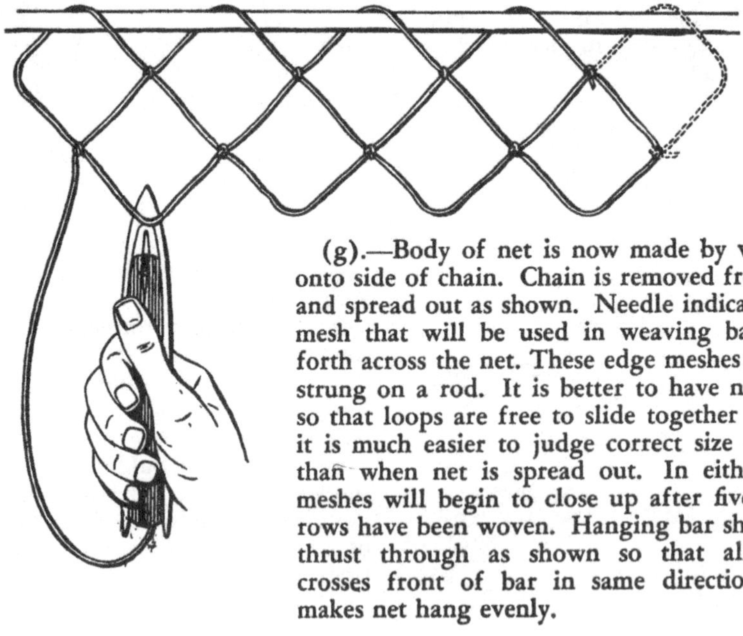

(g).—Body of net is now made by weaving onto side of chain. Chain is removed from nail and spread out as shown. Needle indicates first mesh that will be used in weaving back and forth across the net. These edge meshes may be strung on a rod. It is better to have net hang so that loops are free to slide together because it is much easier to judge correct size of loop than when net is spread out. In either case, meshes will begin to close up after five or six rows have been woven. Hanging bar should be thrust through as shown so that all twine crosses front of bar in same direction. This makes net hang evenly.

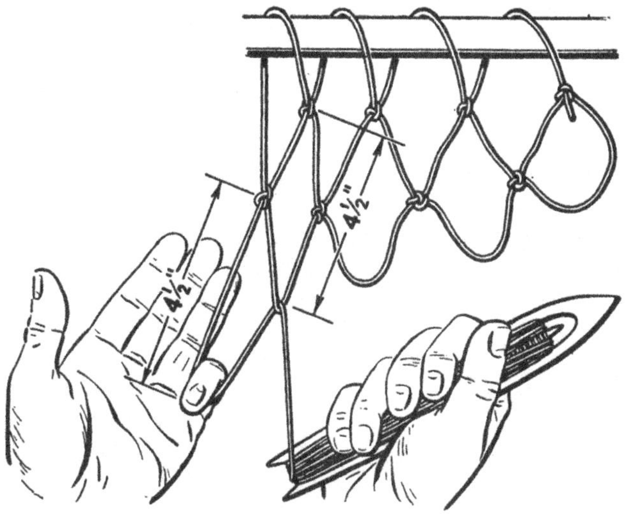

(h).—With chain hung up as described above, twine should lead off from lower left-hand knot. Pass needle up through mesh to right of knot. Hook left little finger in loop. Adjust length of loop so that distance from knot directly above little finger to bottom of loop equals total length of mesh. Complete tie just as was done in making chain. Pick up next loop to right, tie into it, and proceed across to right-hand edge of net. If chain contains an odd total number of meshes, the last mesh on right is strung on rod (dotted lines in fig. 27 (g)) and must be skipped in weaving body of net.

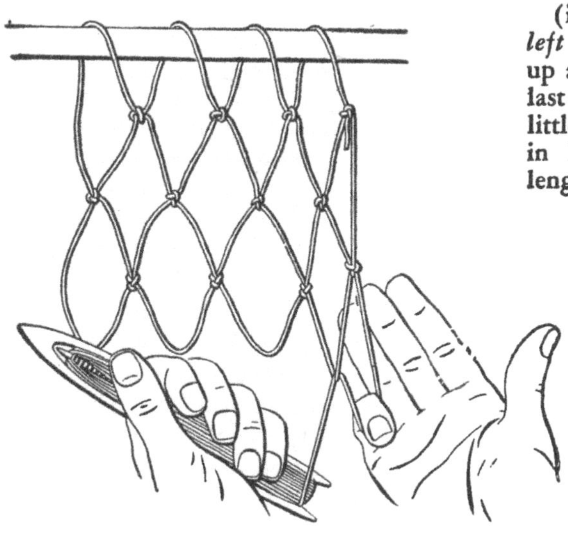

(i).—Change needle to *left* hand, pass point up and to front through last mesh woven, hook little finger of right hand in loop, and adjust to length.

(j).—Throw a loop of twine up to *right* front, pass needle from *left* to *right* behind mesh being tied into, in front of loop hooked on finger, and through loop thrown up to right. Pull knot tight and continue weaving from right to left in same manner until left-hand edge of net is reached. Change needle back to right hand and work back from left to right as in first row of body.

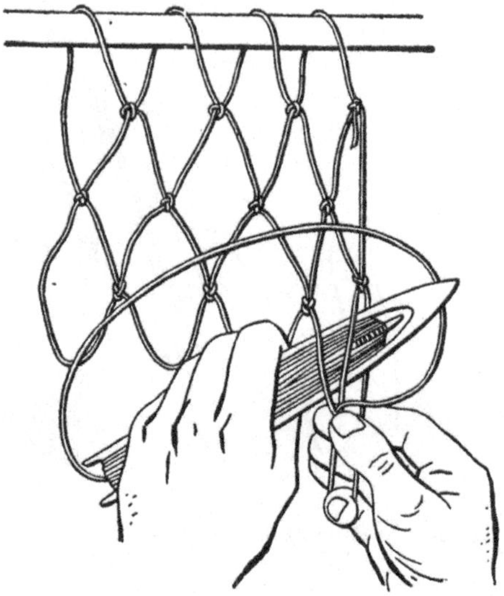

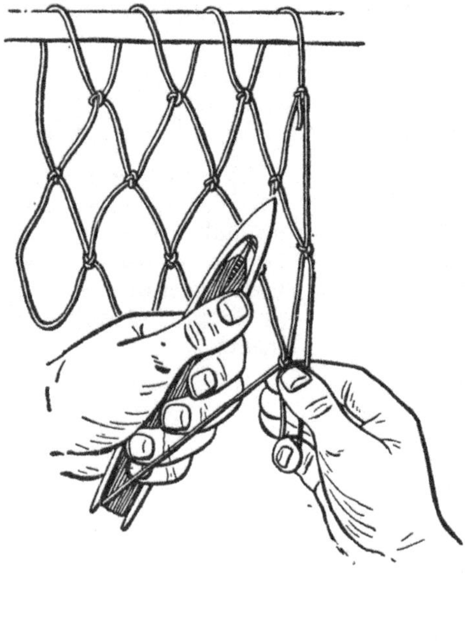

(k).—Continue weaving back and forth until desired length is reached. This length is determined in exactly the same manner as length of chain, by counting mesh if for a patch, or by measurement if for a complete net. Ties must be made as described so that twine will lead directly from one knot to next without crossing twine coming into knot. If trouble is experienced with twine crossing itself when ties are made as described, it indicates that the knots are not being pulled tightly into proper shape. Probably the little finger is not holding loop tightly enough as knot is pulled tight.

SECTION VI. GARNISHING MATERIALS

20. GENERAL.—*a.* Large-mesh twine nets, ungarnished, have no concealing quality in themselves; it is necessary to add either artificial or natural garnishing materials to them before they can be used successfully.

b. The advantages of artificial garnishing materials are: permanence, availability, and prefabrication. Many are mildew proof, fire resistant, and weather resistant.

c. Natural garnishing materials possess advantages not found in the same degree in artificial materials. They are usually found in the vicinity of site, are the same color and texture of local terrain, are effective—when fresh—against detection by aerial infrared photography, and are usually ready to use without processing before application.

d. Occasionally both artificial and natural garnishing may be used together, and this combination may offer the advantages of both types while minimizing the disadvantages of each.

21. CLOTH GARNISHING.—*a. Burlap* (fig. 8).—This material is a loosely woven, coarse-mesh fabric made from the fibers of the jute plant. It tends to fray. It is commonly available and may be issued in bolts 40 inches wide and from 100 to 125 yards long. It is also issued precolored in rolls 2 inches wide by 100 yards long in all standard camouflage colors except white, packed with 10 rolls per bundle. For factory method of cutting and painting bolted fabric into garnishing strips, see paragraph 60*d*.

b. Osnaburg (fig. 9).—This fabric, an unbleached cotton material, is more closely woven than burlap, for which it is a substitute issue. It is available in 40-inch-wide bales made up of bolts of varying lengths. It is

TABLE XXXVIII.—Burlap.

Specifications: Federal specification CCC–B–811 (Untreated). Corps of Engineers Tentative Specification No. T–1212 (Impregnated).

	Color	Stock Number
Rolls 2 inches by 100 yards	Black	49–2400.1–2
	Dark green	49–2400.1–3
	Earth brown	49–2400.1–35
	Earth red	49–2400.1–4
	Earth yellow	49–2400.1–45
	Field drab	49–2400.1–5
	Light green	49–2400.1–55
	Loam	49–2400.1–6
	Olive drab	49–2400.1–65
	Sand	49–2400.1–75
Bolts	Color	Stock Number
40 inches by 100–125 yards. One bale contains 20 bolts or approximately 2,000 yards. (Other widths may be available.)	Sand	not available
	Olive drab	49–2405.4–65
	Natural	27–2590.075–4

Table XXXIX.—Osnaburg.

Specifications: Corps of Engineers Tentative Specification No. T–1212 (Impregnated).

U. S. Army Specification 6–280 (Untreated).

Engineer Board Procurement No. 147 covers rolls of precolored garlands in 2-inch widths.

	Color	Stock Number
Rolls 2 inches by 100 yards	Black	49–2064.05–2
	Earth brown	49–3236.05–25
	Dark green	49–3236.05–3
	Earth red	49–3236.05–4
	Earth yellow	49–3236.05–45
	Field drab	49–3236.05–5
	Light green	49–3236.05–55
	Loam	49–3236.05–6
	Olive drab	49–3236.05–65
	Sand	49–3236.05–75
	Natural	49–2064.2–05
	Color	**Stock Number**
Bolts 40 inches by various lengths (Other widths may be available.)	Natural	27–3014.5–4
	Sand	not available
	Olive drab	not available

also supplied precolored in the standard camouflage colors in 2-inch rolls, 100 yards long, packed 10 rolls per bundle.

c. Prefabricated garlanded wire.—This is proposed as an item of issue. If issued it will probably be in 100-foot rolls of 16-gage wire with fabric garlands (2 in. by 40 in.) firmly attached on 8-inch centers. The free ends of the wire are attached to small wooden blocks so that the outside end of the wire can be found easily. The cloth used is impregnated camouflage fabric in the standard camouflage colors except white. This material is intended primarily for use in blending in the edges of flat-tops and adding cover at the sides of flat-tops. The specification is: Engineer Board Procurement No. 386.

d. Salvaged fabrics.—In the absence of issue camouflage cloth, a good source of fabric for garnishing purposes may be Quartermaster Corps salvage depots. Among the most useful materials which may be found there are: fatigues, old tentage, blankets, raincoats, latrine screens, and ordinary burlap sacks. Judgment should be exercised in choice of available materials, as some are more difficult to cut or paint than others. These materials may also be used effectively for patch garnishing without being cut.

e. Captured fabrics.—Much useful material may be found in captured enemy stores of camouflage material. Also amid the debris of a battlefield may be found civilian curtains, bedding, and other fabrics, which may be used.

f. Locally manufactured fabrics.—Substitutes for issue camouflage cloth may be purchased locally. Care should be taken, when possible, to get fabrics having similar characteristics.

g. Field preparation of garnishing strips.—(1) For coloring small quantities of bolted fabric for garnishing, the following methods may be used. Large-scale production uses the method described in paragraph 60*d*.

(*a*) A 5-gallon bucket, ammunition box, or any watertight container may be converted into a paint vat. The fabric is cut into 5-foot lengths and dipped into the vat so that both sides are completely covered with paint. It is then lifted out, folded in half, placed on a drain trough, and the excess paint removed by rolling (fig. 28). The roller can be made from any small smooth log from which the bark has been removed. The fabric is then wrung out by hand and placed on wires or lines to dry. This primitive

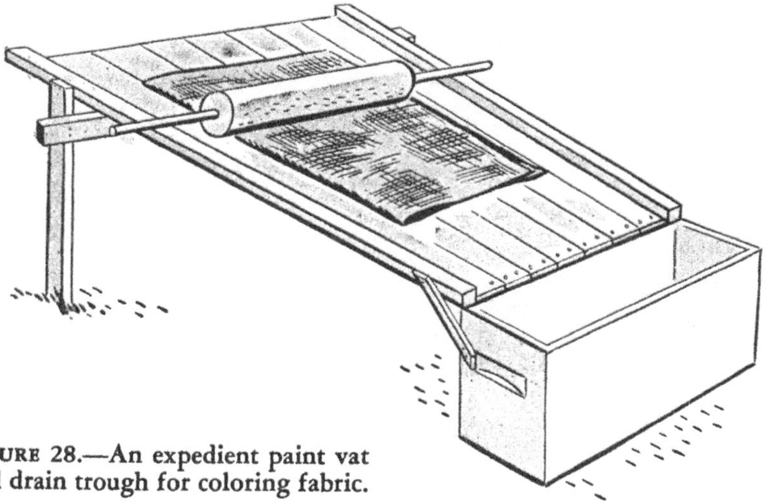

Figure 28.—An expedient paint vat and drain trough for coloring fabric.

method is slow and wasteful of paint. When dry, it is cut into 2-inch widths, as shown in figure 31.

(*b*) A G.I. can do with a rod securely fastened about 4 inches from the bottom and with a clothes wringer attached to the top edge provides a more efficient method of coloring. The bolted cloth is folded in four folds, immersed in the paint vat, passed under the rod, and brought up through the wringer, which removes the excess paint. When dry it is cut into 2-inch widths.

(2) Field-expedient methods for coloring precut rolls of garnishing fabric:

(*a*) The simplest method of coloration is the same as described above for bolted fabrics.

(*b*) A better and faster method employs a clothes wringer (fig. 29). Four rolls of precut garnishing fabric are mounted on an axle supported by a wooden stand. The rolls are carried down into a G.I. can which has a wooden frame wedged into the sides. Nails act as guides. After being threaded through the wringer, the ends of the strips are wrapped around a stick; while one man steadies the can, two others grasp the stick and slowly pull the colored garnishing strips through the wringer. The men walk slowly away from the vat in a straight line for a considerable distance and lay the colored strips on the ground to dry. The strips are then cut off close

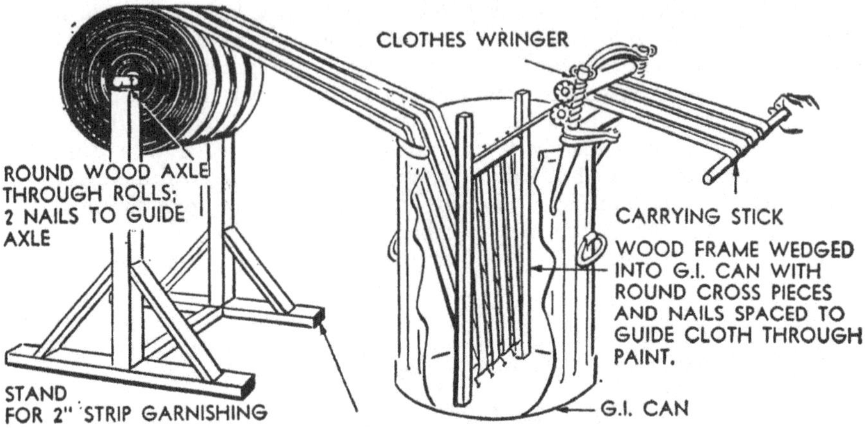

FIGURE 29.—Apparatus for coloring rolls of 2-in. garnishing fabric.

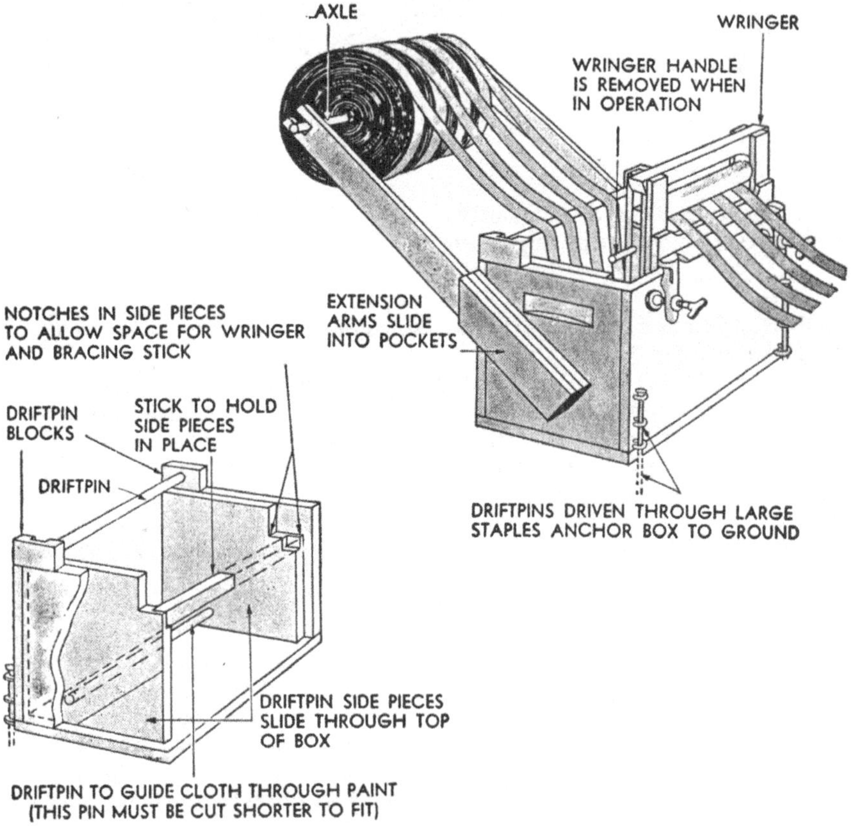

FIGURE 30.—Apparatus for coloring rolls of garnishing fabric, using an ammunition box as vat.

FIGURE 31.—Rolled fabric being chopped by hand into strips, 2-in. wide, for garnishing.

FIGURE 32.—A reel for measuring 5-ft. lengths of garnishing fabric from standard rolls.

to the wringer, the equipment moved slightly to the right or left, and the process repeated.

(*c*) If an ammunition box is available for use as a paint vat, a more compact apparatus can be built by supporting an axle between two extension arms set into pockets on both sides of the paint vat (fig. 30). The extension arms, the two driftpins which guide the cloth through the vat, the driftpin slides, and the bracing stick are removable. The vat is held in place by driftpins which are driven into the ground through large staples or bent spikes placed on the corners. When not in use, the removable parts are placed in the box. Method of operation is the same as in (*b*) above.

(3) Dipped garnishing fabric dries in one to several hours, depending on the weather and the paint employed. It is better to do the drying in the shade. If large quantities of bolted material are to be dipped, a drying rack should be built (par. 60*d* (4)).

(4) Fabrics are cut into the desired sizes as follows:

(*a*) Bolted fabric may be cut into 5-foot lengths by shears, electric cutter, or other means. To convert the 5-foot lengths into strips they may be tightly rolled and chopped into 2-inch widths (fig. 31) or laid flat on a cutting table, folded and cut with the electric cutter.

(*b*) A detail of 5 men using the hatchet method will prepare 900 garlands from 40-inch bolted burlap in 8 man-hours; the same detail using the electric cutter will prepare 900 garlands in 1½ man-hours.

(*c*) After any required re-coloring, the rolls of 2-inch-wide garnishing fabric are wound on a simple reel before cutting them into 5-foot lengths (fig. 32). The reel is pinned to a tree or post with a spike, which serves as an axis. One end of the roll of garnishing fabric is attached to one of the pegs and the reel is turned. The pegs are placed 2½ feet apart, and one complete revolution winds 5 feet of fabric. By counting the number of revolutions it is possible to determine the number of garnishing strips. To cut into 5-foot lengths, tie the reeled strip to prevent it from coming apart, and cut it once.

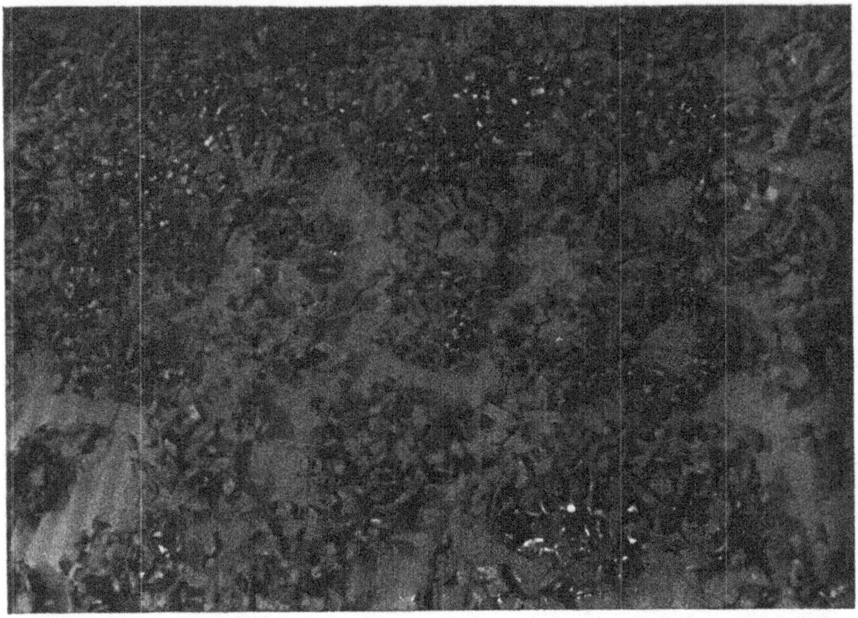

FIGURE 33 ①.—Detail of tin-can garnish applied to roof of a Quonset hut.

②.—A large installation covered with garnished wire netting. Note how the added patches of tin cans blend the cover into the surroundings.

FIGURE 34.

b. Garnishing nets with garlands.—For factory methods of garnishing and rates of production see paragraph 60*b* and *c*.

22. METAL.—*a. Steel wool.*—This loosely matted, finely spun material is particularly useful where a fireproof garnish is required. It has excellent textural quality. It is available prefabricated, bonderized to retard rusting, precolored, and attached to rolls of chicken wire. It requires large amounts of paint to re-color. See table XXXVII for detailed data.

b. Tin cans.—Used tin cans, when mashed or cut into strips and splayed out are a readily available and practical garnishing material of excellent textural quality (fig. 33 ①). The can most widely used in the Army is the No. 10 can (diameter $6\frac{1}{8}$ in., height 7 in.). One can is required to garnish each 1 to 2 square feet of surface. Tin-can garnishing may be employed to add texture to large areas of cloth-garnished overhead nets (fig. 33 ②). A procedure for preparing tin-cans garnishing (fig. 34) follows:

(1) After collecting a suitable number of cans, cut sides into strips approximately 1 inch wide to within 1 inch of the bottom of the can. For construction of a field-expedient cutter, see paragraph 60*b* (1).

(2) Place cans in a fire to remove the light-reflecting lacquer and provide a rough surface on which to paint.

(3) (Optional.) Fabric strips may be crimped to the ends of the tin strips.

(4) The tin strips are bent, some in, some out, to produce an irregular form.

(5) (Optional.) Paint, if required, by dripping, brushing, or spraying. The rusty color of the cans blends with some backgrounds without painting.

(6) (*a*) Attach to chicken wire by clinching two or three of the strips around the mesh—

(*b*) Or nail the can to a flat wooden surface—

(*c*) Or run wire through holes punched in the bottoms of the cans and suspend from an overhead crosswork of single wires.

23. OTHER GARNISHING MATERIALS.—*a. Glass fiber.*—Glass fiber is light in weight, fireproof, and does not employ strategic materials. It is available as a substitute issue for steel wool, is enclosed between two layers of chicken wire, and comes in six standard camouflage colors: light green, dark green, sand, field drab, earth brown, and olive drab. See table XXXVII for detailed data.

b. Paper.—(1) Experimentation is progressing on the development of a tough durable paper substitute for camouflage fabrics for garnishing nets. The advantages of this material are that it is inexpensive, easy to handle, and light in weight. Its main disadvantage is that it has less tensile strength than cloth fabric. It is now available in 2-inch-wide rolls on an experimental basis. Its characteristics indicate that it may be especially suitable for aircraft net sets.

(2) As a field expedient, locally available paper, such as the tough, durable, waterproof paper used in curing concrete roads, may be cut into strips and used as a garnishing material. The danger in the use of ordinary paper garnishing is its inflammability.

c. Processed wood fibers.—Excelsior and other wood fibers have excellent texturing qualities. They are light in weight and are available commercially in many regions. Their only great disadvantage is their inflammability. Other similar woody fibers are coir, which is shredded coconut fiber; tampico, a product of the yucca plant; and redwood fiber, which may be classed as a fire-resistant material. To render them suitable for use on permanent installations, they must be mildew- and fire-proofed, which makes them more expensive than steel or glass wool, and they are less durable.

d. Feathers.—(1) This material has been extensively employed and has proved itself a most useful garnishing material. Sources for this material are: poultry slaughter houses, mattress and pillow factories, and upholstering plants. In remote districts, native headsmen may be able to provide limited

FIGURE 35.—The feather-garnishing machine.

FIGURE 36.—A twine net garnished with feathers.

quantities. It is probable that feather-garnished chicken wire will be a substitute issue for steel-wool garnished netting.

(2) Feathers procured as a garnishing material must be clean, dry, and free from disease and insects. Crushed feathers are acceptable provided not more than 5 percent by weight consist of down separated from the spine. For large-scale coverage, tail and wing feathers are acceptable provided they are crushed in such a manner that sections of the spine do not exceed 3 inches in length. In the field, however, long feathers can be used, in many cases, quite effectively.

(3) Depending on density requirements, the coverage may vary from 7 to 13 square yards per pound of feathers.

(4) (*a*) Feathers are usually made to adhere to wire netting. A machine has been devised to perform the whole operation in a continuous belt system (fig. 35.) For details of construction of the machine see paragraph

FIGURE 37. — Simulated tree, elaborately made of feather - garnished wire netting.

60*g*. A roll of wire is placed at one end, with an axle through the central hole of the roll. The free end of the wire passes under a guide in the bottom of a trough filled with an adhesive, such as bituminous emulsion, and continues under a hopper filled with feathers. The hopper has an agitator at the lower end, which keeps the feathers flowing freely. The feathers fall onto the wire. Two men, one on either side, manipulate a board to spread the feathers over the wire evenly. The wire then passes under a roller to press the feathers firmly onto the adhesive-coated wire. The garnished wire is then re-rolled at the other end. If such equipment is not available, the wire is dipped in a trough of adhesive and the feathers spread by hand. After drying overnight, the garnished wire may be painted.

(*b*) When feathers are applied to twine nets, the nets are dipped into a G.I. can filled two-thirds full of bituminous-emulsion adhesive. To

FIGURE 38.—Simulated rocks made of feather-garnished wire netting.

obtain irregularly garnished edges, the net is held by all four corners in dipping. Only the center of the net thus receives a heavy application of adhesive. The feathers are distributed by hand. The net is rolled and left to dry for two days. After excess feathers are shaken off, the prepared netting is spray-painted if required (fig. 36).

(5) Some feathers may be used without further treatment in areas where their color fits the color of the ground. For example: Plymouth Rock, duck, and goose feathers match gray granite; turkey feathers match slate-colored rock; Rhode Island Red feathers match reddish earth; white feathers match light earth, but appear darker than snow. Otherwise, artificial coloring is normally accomplished by spraying.

(6) The recommended uses of feathers for garnishing are: on chicken wire for permanent flat-tops, false trees (fig. 37), and simulated rock forms (fig. 38); on twine nets; and on walls.

(7) The durability of feather garnishing is good. It offers less resistance to wind than nets garnished with most other materials, allows dry snow to sift through the meshes, and is naturally fire retardant.

FIGURE 39.—An AA gun position is located in this simulated junk yard.

(8) When feather-garnished wire netting is used on flat-tops, the best procedure is to lay out on the ground the required number of garnished strips of chicken wire, and wire together the adjacent sides. The whole is then painted, thus avoiding the striped effect which would result from painting each strip individually. The garnishing at the edge of the flat-top must be thinned or extensions of the netting at the sides must be sloped to the ground.

e. Debris.—In the field, considerable reliance may be placed on the use of local debris as screening or garnishing material. In many cases, debris, such as loose boards, laths, rags, cardboard boxes and military and civilian equipment, gives good concealment (figs. 39 and 40). It requires no renewal and little maintenance.

f. Natural materials, especially cut foliage, sagebrush, straw, seaweed, spanish moss, and growing vines, are excellent garnishing materials. The qualities and preparation of natural materials are discussed in chapter 3.

FIGURE 40.—A dug-in installation effectively camouflaged to resemble a city dump by the use of debris.

SECTION VII. STRUCTURAL MATERIALS

24. GENERAL.—Structural materials are those necessary for building a framework for the attachment, suspension, or support of camouflage materials. They include posts, stakes, wire, wire-rope cable, rope, and twine. These may all be obtained from engineer depots, from commercial sources, or improvised from field-expedient materials.

25. WIRE, CABLE, AND FIELD EXPEDIENTS.—*a. General.*—In some theaters issue wire and cable are extremely scarce; in such instances every effort should be made to limit their use to structures of importance. The steel binding tape used on boxes may be salvaged and used where short lengths are practical. For some uses, locally procurable rope may also be substituted for wire. Barbed wire may be used in the construction of flat-tops in the absence of a more suitable plain wire.

b. Issue sizes.—Wire is issued normally in 16-gage and 10-gage sizes (tables XL and XLI). Other gages available in some engineer depots are:

Gage	*Stock Number*
14	22—8134.5—42
13	22—8134.5—44
12	22—8134.5—46
9	22—8134.5—52
8	22—8134.5—54
6	22—8134.5—58

c. Cable.—Wire-rope cables are usually required where spans in excess of 80 feet are contemplated. To prevent the weight of the cable from causing excessive sag, a cable is used which has the smallest practicable diameter. Turnbuckles are used to take up slack. Clamps are used to fasten auxiliary wire to cables. In emergencies, small lengths of cable may be secured from the motor transportation sections of motorized units. Wire cable may also be made by twisting lengths of 10-gage wire as described in paragraph 25*d*. The most useful sizes of cable for normal camouflage construction have been found to be ¼ inch and ⅜ inch. Numerous types of wire ropes are class IV material; the following items, however, are included in organic camouflage equipment sets: (1) and (2) below are supplementary equipment in Set No. 3 (Camouflage Battalion, Army); (1) below in Set No. 2 (Camouflage Company).

FIGURE 41 ①.—Telegraph wrap with 4 full turns, using pliers.

②.—Square knot tied by hand is as strong as ①, but will elongate from 4 to 8 in. against a 1,000-lb. strain.

③.—Long wrap by hand, 3½ full turns. Although it is the strongest hand wrap, it will unwind at less than a 900-lb. strain.

Table XL.—16-gage wire, steel, carbon, annealed, zinc-coated.

Issue	Rolls, 9,800 feet; weight 100 pounds.
Stock Number	22-8134.5-38.
Average Tensile Strength	250 pounds (high: 300 pounds—low: 200 pounds).
Recommended Uses	To join adjacent edges of nets. To fasten net edges to flat-top frames. To wire in paths. To fasten natural or artificial garnishing to camouflage structures. For general camouflage purposes requiring a light pliable wire, where stretching is not a disadvantage. Not capable of withstanding strong wind pressure or heavy tensions.
Specification	Federal Specification QQ-W-461.

Table XLI.—10-gage wire, steel, carbon, annealed, zinc-coated.

Issue	Rolls, 2,000 feet; weight 100 pounds.
Stock Number	22-8134.5-5.
Average Tensile Strength	1,050 pounds (high: 1,790 pounds—low: 650 pounds)
Recommended Uses	For normal camouflage construction.
Specification	Federal Specification QQ-W-461.

NOTES:

Tensile strength is greatly impaired by kinks that cannot be straightened by hand.

A small cut or nick, such as may be made in tightening a wire by using a clawhammer, reduces its strength at that point.

The strength of various splices indicate the following to be the most efficient in the order in which they are listed (fig. 41):

Telegraph wrap with pliers, 4 full turns.

Square knot tied by hand. Elongation of the splice of from 4 to 8 inches may be expected.

Long wrap by hand, 3½ full turns. Splice will unwind at less than 900 pounds.

Generally in stringing main supporting wires from 2 to 6 strands will be used depending on the length of the span.

Spans 10 feet to 30 feet generally require 1 strand.

Spans 30 feet to 60 feet generally require 4 strands.

Spans 60 feet to 80 feet generally require 6 strands.

Figure 42 ①.—Three- or four-wire cables. Fasten wire to truck towing hook. Loop a continuous length of wire around a stake or tree and back to towing hook until number of strands needed is reached. Secure loose end, and back up truck to tighten loops and even tension in each strand. Insert racking device and twist. The wires are cut near the racking device so that two lengths of cable are provided by this one operation.

②.—Racking device for four-strand cable. Double strands should be racked one full turn per 2 inches.

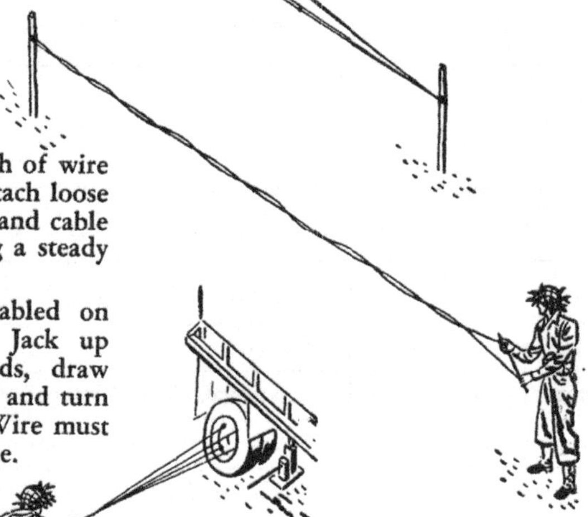

③.—Cut double length of wire and fasten loop end. Attach loose ends of wire to a stake and cable by twisting, maintaining a steady pull to equalize tension.

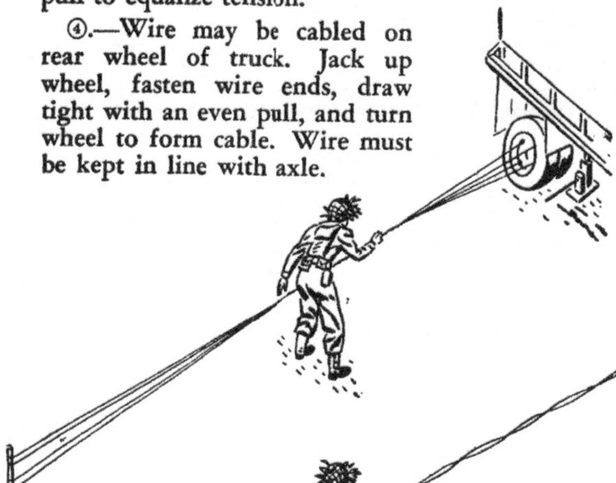

④.—Wire may be cabled on rear wheel of truck. Jack up wheel, fasten wire ends, draw tight with an even pull, and turn wheel to form cable. Wire must be kept in line with axle.

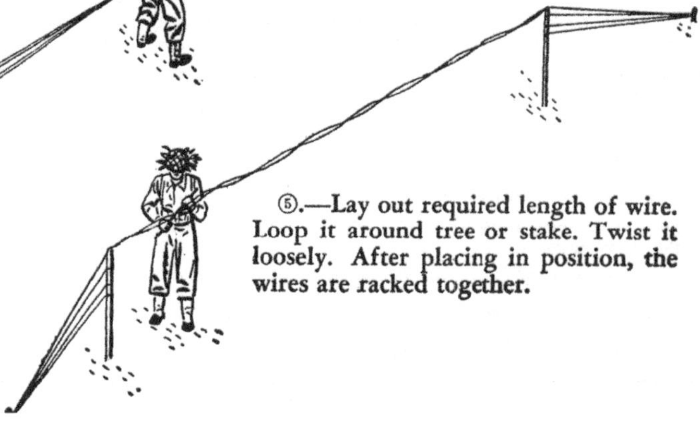

⑤.—Lay out required length of wire. Loop it around tree or stake. Twist it loosely. After placing in position, the wires are racked together.

(1) Rope, wire, steel, tiller, galvanized, fiber core; 6 x 7 stranding 1/4 inch (Stock No. 22-4936.2-03).

(2) Wire, cast steel, galvanized, 6 x 19 stranding 3/8 inch (Stock No. 22-3460.4-04).

d. Field-expedient methods for making wire cable.—When span lengths require stronger wire and steel cable is unavailable, two or more strands of wire may be racked together to form a field-expedient cable. Four simple methods to do this are described in figure 42. Double strands should be racked approximately one full turn per 2 inches.

26. FIBROUS SUPPORTING MATERIALS.—Rope and twine may be used in camouflage construction, especially in temporary installations when wire is not available. The disadvantage of fibrous materials is that they stretch and shrink with changes of humidity. This makes their use less desirable because they require greater maintenance. Many types and sizes of ropes and twine are class II and IV issue, but locally fabricated rope may be procurable in theaters where no issue ropes are stocked. Only marline and tape are included in organic camouflage equipment sets (table XLII). Sisal rope is obtainable locally in some areas and may be used for guying and the support of nets. The sizes in which it may be found vary between 1/8 inch to 1 1/2 inches. The most useful size for camouflage purposes has been found to be 3/8 inch, which has an average breaking strength of approximately 612 pounds.

27. OTHER MATERIALS.—*a. Greenhide rope.*—Cattle stations in Queensland, Australia, frequently make their own rope from untanned, stretched, bullock hides, as follows:

(1) Scrape most of the hair from the hide. Cut it into a strip 3/4 inch to 1 inch wide by starting from the center of the hide and cutting spirally outwards like a flat clock spring. Then soak the strips overnight in water to soften.

(2) To make a 3-strand twisted rope:

(*a*) Take three strips, each about twice the length of the required

TABLE XLII.—Marline and tape.

Issue	Marline, jute, tarred. Two-ply, 10-pound coil with an approximate minimum breaking strength of 110 pounds when new.
Stock Number	21-6520.2-1.
Recommended Use	To mark paths; general utility where suitable.
Specification:	U. S. Army Specification No. 14-3.
Issue	Tape, cotton, general use, bleached; 3/4 inch wide, 500-foot roll.
Stock Number	27-6565-5.
Recommended Use	To mark access paths into installations from existing roads or paths.
Specification:	Federal Specification DDD-T-86.

rope. Tie a movable weight, such as a gasoline can partly filled with stones, to one end of each strip to keep it taut.

(*b*) Make a small slit in each of the free ends and tie them to a winding handle. This may be a metal handle fixed through a tree or post, or even a cart or motor wheel with a hook on the axle.

(*c*) Twist the strips clockwise, and the rope is formed.

(*d*) Allow it to dry for a few days.

(*e*) If time permits, it is preferable to twist each strand separately, counterclockwise, before all three are twisted together clockwise. A much better rope is produced in this way.

(*f*) The two ends of the finished rope should be seized or whipped.

b. Bamboo rope.—Another expedient for rope is to split green bamboo of the hollow type into small sections and split each section several more times. Plait three of these long thin strips together. To complete the rope, take four of the resultant 3-ply ropes and plait together.

c. Creepers.—Similarly, creepers may be woven into ropes.

28. POSTS, STAKES, AND OTHER STRUCTURAL MATERIALS.—Normally in the field, posts and stakes are procured locally. They may be cut on the site from existing small trees or salvaged from abandoned civilian structures. In regions where bamboo is a native growth, it is suitable for light construction work. Barbed wire pickets and round steel stakes may be used in place of wooden stakes, and in some types of ground are more efficient.

CHAPTER 3

NATURAL MATERIALS

		PARAGRAPHS
Section I.	General	29–34
II.	Ground Patterns and Natural Materials.	35–38
III.	Overhead Cover and Natural Materials.	39–45

SECTION I. GENERAL

29. INTRODUCTION.—In all camouflage planning, natural materials should be used whenever practicable because of their availability, their adaptability to many uses, and their excellent camouflage properties. Natural materials may provide two-dimensional camouflage having only length and width or three-dimensional camouflage having length, width, and height. In order to use live vegetation properly, camouflage supervisors should be familiar with the common characteristics of living plant materials and the techniques of planting, fertilizing, transplanting, and processing vegetation. However, for best results, the advice of experts should be obtained; this is especially important in overseas theaters. Possible sources of competent advice (in the absence of Corps of Engineer specialists) and of raw materials include:

a. Local universities.

b. Governmental agencies—agricultural and forestry branches.

c. Nurseries.

d. Local garden clubs or similar organizations.

e. Local farmers.

30. TYPES OF NATURAL MATERIALS.—*a*. There are three types of natural materials: live vegetation, cut vegetation, and the inert covering of the earth, such as soil, rocks, and gravel.

b. Natural live vegetation, except coniferous vegetation, possess infrared reflectance, is available almost universally—except in desert regions and extreme arctic regions such as Greenland—is permanent, improving with continued growth, and requires a minimum of maintenance. However, the planting or transplanting of plant material requires time, skilled labor, and an adequate water supply—especially in the early stages; if large trees are to be transplanted, heavy equipment is necessary.

c. Cut vegetation is available in most regions in adequate quantities. However, it requires maintenance, loses the property of infrared reflectance, and is inflammable.

d. Inert substances which are useful are limited. Those of value, such as colored soil and crushed rock, provide either a change of color or texture or both and are normally available in large quantities.

31. DESIRABLE CHARACTERISTICS OF LIVE VEGETATION FOR CAMOUFLAGE PURPOSES.—For each planting project the key to proper

selection of plants will be furnished by an investigation of what is already succeeding well in the locality. Plants may be native or thoroughly naturalized; those varieties found thriving under adverse conditions are likely to be especially desirable. Only plants capable of growing under conditions at the actual site of planting should be used; fertility of soil, wetness or dryness of ground, and degree of exposure to wind and salt air must be taken into account. The following characteristics are desirable:

a. Ability to survive; this is dependent on—

(1) A normal, vigorous root system.

(2) The use of native or naturalized plants rather than imported species.

b. Rapid establishment and growth without undue care.

c. Adequate covering capacity of the foliage.

d. Freedom from disease, insect pests, and abrasions of the bark.

32 TYPES OF VEGETATION.—The kinds of vegetation useful for camouflage purposes include:

a. Legumes and low-growing plants.

b. Grasses.

c. Vines—deciduous or evergreen.

(1) Horizontal-growing varieties.

(2) Vertical-growing—twining or clinging on supporting objects.

d. Shrubs—deciduous, evergreen, or broadleaf.

e. Trees—deciduous, evergreen, or broadleaf.

33. USE OF VEGETATION IN CAMOUFLAGE.—The type of vegetation and the method of using it are largely determined by the camouflage plan. Foliage may be used for two different purposes: to ground-pattern the earth surface by changing its color or texture (fig. 43) or to provide overhead cover. Either method, when properly employed, blends the treated area with its surroundings. For data on restrictions occasioned by operational requirements—especially as related to airfields—refer to Engineering Manual, chapter XXII, paragraphs 22 to 32. The techniques involved are described in sections II and III of this chapter.

34. LIMITATIONS OF LIVE VEGETATION IN CAMOUFLAGE.—At least five factors limit the range of choice of vegetation and the practical methods of employment:

(1) Can the job be completed in time to be effective?

(2) Is sufficient plant material available to provide effective camouflage, is it adaptable to growth at site?

(3) Is sufficient trained labor available?

(4) Is supply of water sufficient?

(5) Are tools, supplies, and equipment available?

SECTION II. GROUND PATTERNS AND NATURAL MATERIALS

35. GENERAL.—A ground pattern is made by changing the texture or color of one or more parts of the ground surface according to a predetermined plan. When seen from the air, it must conform to the natural pattern of its surroundings; otherwise, it will invite enemy suspicion. Whole areas may be given a homogeneous tone, or narrow strips simulating hedges, paths, ditches, or other common terrain lines may be made. Ground patterning is used primarily for the concealment of airfields; secondarily, for supply points,

FIGURE 43.—This two-dimensional pattern, applied to a runway, gives effective variation in color and texture.

other areas of large size, and small installations. This section suggests some of the many materials and ways of using them. Many other materials may be used and new techniques of application devised.

36. METHODS OF GROUND PATTERNING.—These are divided into two general classes, depending on the result to be accomplished:

a. Two-dimensional ground patterning produces a change in color and, to some degree, in texture over a relatively flat area (fig. 43).

b. Three-dimensional ground patterning introduces the element of height, thus providing shadow and stronger contrasts of texture and color (fig. 44).

37. TWO-DIMENSIONAL GROUND PATTERNING.—*a. Spreading materials over a surface.*—Two types of natural materials are used: cut vegeta-

FIGURE 44.—Three-dimensional artificial trees used to add realism to an airdrome camouflage scheme.

tion or inert materials, such as soil, gravel, small rocks, and other loose material.

(1) *Plant materials,* such as heavily leafed branches, dead branches, saplings, corn stalks, tumbleweed, sagebrush, reeds, cattails, kelp, spanish moss, chaff, vines, tall-growing field grasses, are among the available materials.

(2) *Inert materials* are found in most localities in a considerable range of native earths of varying color and texture. Properly used, they change the ground pattern according to the camouflage scheme.

(3) The choice of material and the method of distribution depends on availability of material, the color and texture desired, climatic conditions—especially wind and rain—and the purpose for which the area is used. Strong winds will blow away light materials, and heavy rains will wash away thin layers of sand or earth used as covering materials. Fine materials may form dust clouds which are injurious to equipment such as aircraft. Normally materials are spread more or less evenly over the surface to be treated. However, to secure the effect of a cultivated truck garden, individual tumbleweed may be arranged in rows and staked to the ground. They may be colored green by dipping or spray painting. Mulching is a good method to tone-down scarred earth, assist dust control, and aid in the establishment of turf plants. Peat, muck, and straw are common materials used for this purpose.

b. Scarification is disturbing or breaking up the surface of the ground. By varying the method used, a considerable range of differently textured surfaces is obtained. Scarification is a quick method of ground patterning and is normally suited only to areas where farming or the clearing of ground is a common activity. This method should not be used where dust may be a menace and normally will be used only as a temporary measure to be replaced by seeding for grass. Engineer equipment, locally available farm or road-building machines, or improvised devices may be used to produce the desired results. By varying the direction of drag, different patterns may be achieved from the same equipment in adjacent areas. Where necessary, this method may be used to obscure tracks and scars, especially those resulting from construction work. Methods of scarification include:

(1) *Rooting,* which produces a rough, coarse texture without geometrical pattern on any surface, including soil containing rocks and boulders. This is accomplished very rapidly by an issue rooter hauled by a track-type tractor.

(2) *Plowing,* which produces a coarse texture with a distinct geometrical pattern over any surface except hard ground. It is accomplished by standard issue equipment (such as a tractor plow, disk plow, or gang plow) or by farming equipment.

(3) *Scarifying,* which produces a coarse texture on most surfaces. It is ineffective where heavy sod occurs. The scarifying teeth on issue blade graders or rollers are used for this purpose.

(4) *Cultivating,* which produces a medium coarse texture over bare ground and very light sod. Issue chisel cultivators or similar farming equipment are used.

(5) *Harrowing,* which produces a medium fine texture and is recommended over plowed surfaces or bare ground which is not too hard. Issue equipment for this purpose includes the disk and spike-tooth harrow. A drag can be improvised to serve the same purpose by driving spikes or drift-

FIGURE 45.—A decoy runway being made by burning grass. Note main fire and backfires.

pins through a timber frame which is sufficiently weighted to bear hard against the soil.

(6) *Smoothing or blading,* which produces a smooth, textureless surface over bare ground. This is accomplished by the use of the blade on standard issue motor or pull-type graders.

c. Spot and pattern irrigation is a means of developing patterns in turf in dry areas or during dry periods. This method is recommended for small areas, where it is particularly effective. If ample water supply and water distributing equipment are available, its use over large areas is possible. When watering, avoid saturation; saturation reduces the bearing capacity of the soil. To prevent too rapid drying, watering should be done in the late afternoon or at night.

(1) In farming country, the watering of predetermined areas results in creating the effect of irrigated fields and crops. During a summer drought, watering confined to areas of an all-over turf field alters the uniform drab color of the turf and simulates patches of alfalfa, clover, and other drought-tolerant field crops which remain green while turf itself is ordinarily dormant.

(2) If water-distributing equipment which is standard Corps of Engineers issue is unobtainable, other equipment may be improvised or borrowed from local farmers and other civilian sources.

(3) 27,000 gallons of water are required per acre of ground to provide the amount of water equivalent to a natural rainfall of 1 inch.

d. Burning dry grass offers one of the quickest methods of runway simulation. The darkened area so made normally lasts only a few weeks. No hard-and-fast procedure can be outlined for this method, but certain precautions must be followed, not only to obtain the intended result but to do it safely. The following sequence of procedure is generally applicable:

(1) The decoy runway or other area is first oriented, laid out, and staked.

FIGURE 46.—A decoy runway in the process of being burnt off.

(2) The boundaries of the runway or area are then ditched or plowed so that a strip of soil is left bare of vegetation as a fire gap and to assure that the edges of the area are straight.

(3) Adequate precautionary measures for fire control are taken. These include:

(*a*) Wetting boundaries of the area to be burned by standard army water sprayer.

(*b*) Providing trucks loaded with sand, brooms, rakes, shovels, or mobile fire-fighting equipment at the area.

(*c*) Providing personnel to control the fire.

(*d*) Starting fires only in winds of lesser velocity than approximately 10 miles per hour (At this velocity, leaves and small twigs are kept in constant motion; when branches begin to sway, the wind is greater than 12 miles an hour.)

(4) Main fire is set on the up-wind side. At the same time, a smaller backfire is set on the down-wind side.

(5) Level brush not flattened by fire to avoid irregularity of surface texture.

(6) If burned area is not sufficiently dark, cutback asphalt or bituminous emulsion may be applied, or a natural black material, such as cinders, may be spread over any light colored areas to darken them.

e. Grass planting provides a permanent and effective tone down or two-dimensional ground patterning on bare or sparsely covered earth. It is also useful in preventing the formation of dust clouds which are visible for great distances. The discussion that follows is a condensed guide based on the three methods of grass planting: seeding, sodding, sprigging. Availability of grass seed, tools, and equipment depends on local conditions. Essential equipment may be improvised or converted from other primary uses for planting grass. Effective patterns to create the appearance of separate fields and diversified methods of culture are possible when strongly contrasting species are planted in adjacent areas as, for example, alsike clover and rye grass, or cowpeas and grass. The most realistic ground patterns are made by combining planting with a planned system of control, using mowing, irrigation, and fertilizing.

(1) Selection of grasses is limited to native or naturalized species suited to a particular locality. Attempts to use species outside this range almost invariably meet with failure.

(2) Planting seasons vary considerably. Generally, in climates comparable to those found in the United States, the following statements apply:

(*a*) In cool humid regions, seeding of permanent species is done preferably in late summer or autumn. Temporary planting is done in spring or summer. Planting grass-seed mixtures is the common practice.

(*b*) In warm humid regions, seeding is done during spring or summer months. Planting unmixed seed is the usual practice.

(*c*) In dry regions, sodding or sprigging is done preferably at the beginning of the normal rainy season.

(3) The best type of soil is a sandy loam. Topsoil may be saved and redistributed after grading. Some subsoils need no treatment to support the growth of grass. Sandy soil, however, should be treated by the addition of a stabilizing or binding layer of clay and peat or muck. On the other hand, sand may be added to heavy clay soil.

(4) Drainage is extremely important in the establishment and maintenance of turf fields. It should be remembered that water drains more slowly from turf than from pavements.

(5) Preparation of the seedbed is influenced by the purpose for which turf is planted, size of area, and soil type; these have been discussed above. After careful consideration of such factors, the procedure is normally as follows:

(*a*) Grade the area with standard issue or commercial graders.

(*b*) On large areas, such as fighter airfields, the preparation of soil for planting can sometimes be accomplished in one operation by the use of an issue or commercial farm disk. This operation may leave the seedbed in a rather rough condition, but because of the dust problem the rougher seedbed in dry areas is superior to one on which the soil has been finely pulverized. In tightly packed clay soils it may be necessary to use plows or other tillage methods. If fertilizer is applied before the final disking, the same operation can be used to work the fertilizer into the soil and to break up the soil properly for planting. The fertilizer should not be disked into the soil below the level at which most of the grass roots will grow. On airfields, most farm disks are not suitable unless followed by a drag or harrow because they are constructed to throw the soil two ways. The use of this type of disk leaves alternate ridges and furrows, and, unless leveled by some type of drag, they cause a great deal of vibration in aircraft traveling across the field.

(*c*) On smaller areas, small disk plows or tillers may be used. On very small areas, spade by hand to a depth of 3 to 6 inches, depending on the soil type. Hand raking is employed to smooth and pulverize the soil sufficiently to make it suitable for planting.

(6) Fertilizing requirements of grasses vary in relation to the nutrient content contained in a particular soil. Whenever possible, soil tests should first be made through local sources. Soil-test interpretation for camouflage purposes is concerned primarily with cover and color, and not with productivity of crops planted.
The following basic practices can be considered only as a guide.

(*a*) *Proportions of constituents.*—Fertilizer mixtures containing approximately half as much phosphoric acid as nitrogen and still less potash give excellent results in most climatic and soil conditions.

(*b*) *Variations in proportions.*—Inorganic fertilizers of 10-6-4 and similar proportions should be applied at the rate of not less than 400 pounds to the acre. Content of commercial fertilizers is designated by numbers which indicate the percentage of their nutrients: 10-6-4 is ten percent nitrogen, six percent phosphoric acid and four percent potash. If fertilizers with different percentages of nitrogen are used, the rate should be modified so that corresponding amounts of nitrogen are applied. On areas such as roadsides or shoulders of runways, where fertility is extremely low, it is advisable to apply two or even three times this amount during the preparation of the original seedbed. When phosphorus or potassium is lacking apply larger amounts of phosphoric acid and potash than are present in a 10-6-4 fertilizer.

(*c*) *Fertilizing as maintenance.*—After turf has become established it usually will be necessary to apply fertilizer occasionally to maintain a heavy stand. The frequency and rate of application depend on the fertility of the soil and the quality of turf required. It is not safe to apply inorganic fertilizers at rates heavier than 40 pounds of nitrogen to the acre (400 pounds of 10-6-4 or 200 pounds of sulfate of ammonia) because of the danger of burning the foliage. Also, to avoid serious burning of the grass, apply only when the blades of grass are dry. Do not apply inorganic fertilizers to young seedling grass.

(*d*) *Seasons for application.*—Inorganic fertilizers are applied to turf just prior to the time of most rapid growth of the grass concerned. In cool humid regions, applications in late summer or early autumn encourage maximum growth of the turf grasses after the annual grasses and weeds have been killed by the first light frosts. In these regions, fertilizer applications, if made in the spring, should be made very early. In warm humid regions summer-growing grasses are fertilized in the spring to encourage the spring and summer growth. On light sandy soils in either region, where the nutrients in inorganic fertilizers leach rapidly, make several small applications during the growing season rather than one large application.

(*e*) *Organic and inorganic fertilizers compared.*—For most conditions the inorganic mixtures are more satisfactory than the organic from the standpoint of cost as well as quick reaction on grass growth. Where large local supplies of organic waste materials such as activated sludge are available, advantage should be taken of such materials; but it should be remembered that while their effect may be more lasting than is that of inorganic mixtures, the response of the grass will not be as quick. The chemical analysis of such materials should be determined, and they should be applied at rates which furnish at least 40 pounds of nitrogen to an acre.

(*f*) *Lime.*—Finely ground limestone is commonly used on highly acid soils. Unless soil is very strongly acid, the rate need not exceed ½ to 1 ton per acre. If hydrated lime is used, it should not be applied within a period of 2 weeks of the time of application of fertilizers containing ammonium salts.

(7) *Planting grass* is accomplished by planting grass seed, planting fragments of actively growing grass, called sprigging, or by transplanting grass sod. Sodding and sprigging are the quickest methods of obtaining a tough wear-resisting turf.

Figure 47.—Drill seeding a section of an airfield. A single drill may be used nearly as effectively.

(*a*) Good seeding depends on even distribution of seed, depth of planting, and rate and time of seeding.

1. Broacast seeding is done by hand or by mechanical seeders. Even distribution is accomplished by sowing half of the seed over the entire area in one direction and the remaining half crosswise. A rake, drag, or harrow covers the seeds to the desired depth.

2. Drill seeding is limited to the availability of drill equipment. Some drills have the advantage of distributing seed and fertilizing in the same operation (fig. 47).

(*b*) Sprigs are fragments of grass producing runners either above or below the ground. Each runner will produce roots when covered with soil and provided with adequate moisture. Only grasses producing such runners are used for sprig planting. Sprigs must be kept moist and be covered immediately after planting to prevent drying out.

1. Spot-sprigging is the planting of sprigs at regular intervals in rows. This requires the least material, but takes the longest period of time for a dense cover of turf.

2. Row-sprigging is the placing of sprigs in furrows and covering with a plow, drag, or hand rake (fig. 48), or by forcing the sprigs into the ground by hand. Effective sprig-planting machines have been developed by the Corps of Engineers. This method uses more planting material than spot-sprigging.

3. Broadcast-sprigging is the distribution of sprigs over an area by hand or machinery. In small areas, the sprigs are scattered by hand, rolled, and then covered by a thin top dressing of soil. In large areas, sprigs may be distributed by a manure spreader or woodchip spreader.

(*c*) Thin sod is easier to move than thick sod. Thin sod cut in 1-inch or even ¾-inch depths has more surface roots exposed and therefore becomes anchored more quickly, but it dries out very rapidly. Thin sod should not be used unless plenty of moisture is assured either by rainfall or

FIGURE 48.—Native labor engaged in planting grass on an airdrome by the row-sprigging method.

through available watering facilities. Sod cut in 2- to 4-inch depths will not dry out as quickly, but requires more labor to move and a longer time to become anchored.

1. Spot sodding is the planting of small blocks of sod at regular intervals when grasses used are of a rapidly spreading species. It is used chiefly in areas subject to periods of drought.

2. Solid sodding is the complete covering of an area by strips or blocks of sod. This method requires the most material and largest amount of labor.

3. Strip sodding is used successfully in many cases on slopes which are too steep for other planting methods to succeed (fig. 49). Spaces between strips may be seeded or sprigged.

f. Mowing produces effective changes in the color and texture of grassed areas (fig. 50). Mowing is especially suitable for the camouflage of any area. Differences in tonal values are marked during periods of maximum growth, but during dormant seasons differences between mowed areas, varying from 2 to 6 inches, are not apparent at distances of more than 1 or 2 miles and at altitudes above 2,000 feet. To maintain a distinct difference at all seasons, turf in certain areas are allowed to mature without mowing, while other areas are regularly mowed at heights of 3 to 6 inches. Careful and regular maintenance is essential. A few of the more common effects obtainable are described below:

(1) Ditches can be simulated by mowing or hand cutting tall grass in lanes from 4 to 6 feet wide (fig. 52). The difference in texture between the mowed and the unmowed portions, and the resulting cast shadow, produce the effect of ditches to the air observer.

(2) Simulation of field cultivation is obtained by mowing adjacent areas in various directions and at different heights.

FIGURE 49.—A revetment being strip-sodded by native labor.

FIGURE 50.—A distinct pattern produced by cutting strips of various heights in a turf field.

FIGURE 51.—Grass landing field in foreground has been blended with the surrounding ground pattern by painted hedges and controlled mowing.

(3) Four or more pattern changes may be accomplished by combining fertilizing with varying heights of cut. Grass mowed at 2 to 3 inches may be fertilized to produce one pattern, while an unfertilized portion of the same cut will produce a pattern of different texture and color. Grass mowed at 5 to 6 inches will produce a coarser texture and a darker ground color than low mowing.

g. Fertilizing applied to turf produces increased growth, greater density, and thereby a darker tone. By regulated fertilizing of different areas of a field, several grouna colors and textures are produced (fig. 53①). Fertilizing for this type of ground patterning is most effective in the spring or autumn seasons.

(1) The effects possible by using controlled fertilizing are most deceptive at low altitudes. At high altitudes and oblique angles the patterns are less noticeable and may be lost entirely.

(*a*) Hedgerows are simulated by applying fertilizer in irregular

FIGURE 52.—A ditch simulated by cutting a lane through high standing grass.

Figure 53 ①.—Controlled fertilizing produced the darker strip of grass in the center of this photograph.

strips; when applied in curves the contours of rolling terrain may be simulated.

(*b*) Row crops are simulated with the use of fertilizer drills (farming equipment) by closing some of the regular spaced openings in the drill.

(*c*) Ripened grain or other brown vegetation may be simulated by applying fertilizers in excess or during unfavorable seasons. These are temporary burns and do not destroy soil or vegetation.

(*d*) Orchard plantings may be simulated at favorable angles by uniform application of high-nitrogen fertilizers to regularly spaced circular spots arranged either in straight rows or along false contour lines to suggest hilly terrain.

(2) Recommended fertilizers are:

(*a*) Sulphate of ammonia and nitrate of soda, or prepared fertilizers containing them, for quick stimulants.

(*b*) High-nitrogen fertilizers.

(*c*) Use above fertilizers at a rate which provides approximately 40

②.—Chemical killing agents produced the lighter-colored rectangle of turf.

pounds of nitrogen to the acre. Increase this rate on very poor soils to give the desired result.

(*d*) These same fertilizers are useful in causing burns, and subsequent browning, by excessive applications. Burns may also be produced by spraying the turf with 1 to 5 pounds of sulphate of ammonia per 1,000 square feet. The heavier rates of application are needed in cool dry weather. In hot humid weather maximum burning is effected and rates of application must be kept at the minimum to avoid permanent damage to turf. In case spraying equipment is not available, maximum burning of grass is accomplished by spreading fertilizers on grass when it is wet with dew.

h. Chemical killing agents produce a peculiar burning and discoloration on turf, which provides another but somewhat limited method of ground patterning (fig. 53 ②). The injury to the turf is only temporary, but must be carefully controlled by varying the rate of application according to season, soil conditions, and type of vegetation.

(1) Suggested uses of killing agents are:

(*a*) Spraying one or more sections of a turf airfield with one or more chemical killing agents to produce areas of different colors, giving the appearance of separate fields.

(*b*) Simulating row crops at the appropriate season by closing some of the openings on a boom sprayer to give proper spacing.

(*c*) Spraying strips or bands of turf 75 to 100 feet wide in curved lines to conform to contour lines in the vicinity; this makes a level surface appear to be a hill or a gully.

(2) Notes on the use of killing agents: (*a*) Spraying, using issue or farm equipment, produces best results.

(*b*) In burning vegetation with killing agents, the solution should be as concentrated as possible. The most severe burns are made during hot humid weather, so rates of application can be lower at such times. This accomplishes a maximum of burning with the smallest amount of water.

(*c*) Some chemicals are extremely poisonous internally and externally and should be handled accordingly.

(3) Types of killing agents, application, and coloring properties follow:

(*a*) Sodium arsenite and other arsenicals produce a straw color, which closely resembles ripe grain and similar grasses. They may be used at rates of 10 to 40 pounds to the acre. This killing agent is extremely poisonous and should be carefully handled.

(*b*) Iron sulphate blackens grass and has possibilities for creating shadow patterns on turf. From 50 to 250 pounds are applied per acre.

(*c*) Ammonium thiocyanate has a bleaching effect on turf, lasting from several days to a month or more. The nitrogen it contains then becomes available and stimulates the grass to produce a darker, more vigorous growth. From 50 to 250 pounds are applied per acre, depending on growth and weather conditions.

(*d*) Aniline dyes have been used on turf to produce a dark green color. A common mixture contains approximately two-fifths malachite green and three-fifths auramine, with a trace of crystal violet.

(*e*) Ammonium thioarsenite and ammonium sulphate may also be used for this purpose.

(4) Since spraying equipment varies, each machine should be calibrated by spraying a measured amount of water on turf. Sufficient water should be applied to cover thoroughly the foliage without dripping. The requirement per acre for each type of spray outfit can be determined by measuring the area covered and the amount of water necessary. The amount of killing agent required for the area to be sprayed can then be dissolved in the amount of water needed.

38. THREE-DIMENSIONAL GROUND PATTERNING is a logical development from two-dimensional patterning and differs from it only in that it involves considerable textural change (fig. 54). It may be accomplished by:

a. Spotting movable three-dimensional material, such as haystacks, corn stalks, and other natural material, on a flat surface so its appearance is materially altered.

b. Transplanting bushes and shrubs to produce three-dimensional ground patterning. The method of transplanting is fully described in paragraph 44.

c. Planting vines to produce a three-dimensional pattern. Since vines in temperate zones are slow in growth compared with those in tropical zones, consideration must be given to the time necessary to produce adequate growth. Planting vines in permanent and semipermanent installations is covered in paragraph 43.

FIGURE 54.—An aircraft hide located in house of a simulated farm. Farm is complete with orchard, kitchen garden, and buildings necessary to accomplish the intended illusion. Path goes to false woodpile where gasoline drums are stored and to outhouse which connects with an underground briefing room. Chicken coop at lower right conceals tents.

SECTION III. OVERHEAD COVER AND NATURAL MATERIALS

39. GENERAL.—Overhead cover may be created by attaching natural materials to a framework or by planting and transplanting tall-growing or wide-spreading vegetation. Overhead concealment serves primarily to screen an installation from aerial observation.

40. NATURAL MATERIALS ON WIRE FLAT-TOPS OR OVERHEAD SCREENS.—

a. Cut foliage is excellent for overhead screening. Care must be taken to place green vegetation in a natural position because the underside of leaves photographs lighter. Lengths of wire netting are placed side by side without overlapping and without fastening adjacent edges. Men working in the open spaces between the lengths stick branches in the wire netting so that the butt ends project down 8 to 12 inches. The butt ends are then wired to hold the branches upright (fig. 55 ①). Another method is to thrust cut branches through two separate layers of wire netting spaced about one foot apart (fig. 55 ②). For most useful types of foliage, see tables XLIII to XLV; for foliage preservative methods, see paragraph 42.

b. Cut and tufted bunches of straw or grass, wired or tied together and with the ends dipped in tar, afford a covering material of good color and texture. They may be forced into nails projecting from strips of wood, inserted into the meshes of wire netting, or suspended by wires (fig. 56 ①). Loose hay or straw may be thatched over a framework of bent saplings covered with wire netting to simulate haystacks or native huts (fig. 56 ②). One method of fastening the material is to cover it with another layer of ungarnished wire netting. The chief disadvantage of dry grasses and straw is their extreme inflammability.

c. Debris, supported by netting, provides a covering material of excellent camouflage value.

FIGURE 55 ①.—Cut foliage inserted in a low flat-top. Butts are anchored with wire.

②.—Cut foliage inserted through two layers of wire netting.

FIGURE 56 ①.—A structure camouflaged with the aid of cut and tufted bunches of local grass affixed to wooden slats.

②.—A simulated haystack conceals a building occupied by troops.

FIGURE 57.—Overhead cover provided by vines growing from tin cans fastened to a hastily erected flat-top.

d. Vines and tubers, such as sweet potatoes, yams, and Chinese potatoes, may be placed in water-filled tin cans secured to wire netting. Cans are painted olive drab or smeared with local mud. A few leaves or a piece of burlap floating on the water in the cans kills reflections. Train vines to grow over the netting (fig. 57).

e. Wetted moss, especially Spanish moss, may be grown on netting. It attaches itself to the netting and sustains life from the air. Green when damp, it changes to gray-green when dry.

41. CHOOSING CUT FOLIAGE.—*a.* The material selected determines whether camouflage will last for several days or turn brown and die within a few hours.

b. The following principles apply to all plants:

(1) Foliage from plants which wither rapidly never should be cut for camouflage purposes; They include: willow, aspen, butternut, walnut, feathery leaved acacia, and American ash.

(2) Choose branches which have been growing in direct sunlight. These stay green much longer than branches from the same tree grown in the shade. The reverse is true in the tropics.

(3) Choose foliage with leaves that feel tough to the fingers.

(4) Cut foliage at night, early in the morning, during a rainstorm, or

in heavy fog. It stays green longer than foliage cut in direct sunlight. This is especially true if branches are to be supplied with water.

(5) When cut foliage is placed in the same position as growing branches it stays green longer; if leaves are turned upward, un unnatural appearance results; and the tender undersides of leaves turn brown quickly when exposed to direct sunlight, thereby losing their value for camouflage.

(6) Recutting stems as soon as the first signs of wilting appear materially lengthens the life of cut foliage supplied with water. When practical make cut under water.

c. Notes on various species follow: (1) Softwoods and hardwoods are the two principal kinds found outside the tropics. Table XLIII lists the lasting qualities of both.

(*a*) Softwoods, known also as conifers, are evergreen trees with needle-like leaves. The group includes pines, spruce, fir, larch, hemlock, cedar, juniper, cypress, and many others. Larch and southern bald cypress, however, lose their needles in winter. Foliage of softwoods lasts much longer than hardwood foliage and should be used wherever available and appropriate. Pines and spruces last from one to two weeks. Hemlock and birches are the poorest of the softwoods for camouflage and last only two to four days.

(*b*) Hardwoods, also known as deciduous trees, are broadleafed. Oaks, maples, and all other trees and shrubs except softwoods (see par. (*a*) above) and such tropical plants as palms and bamboos, are hardwoods. Although each species of hardwood has different lasting qualities, different members of the same species behave similarly. For example, the foliage of gray, paper, yellow, and black birches stays green for about the same period of time, and for camouflage purposes it is only necessary to know that birches as a group last 3 to 4 hours in direct sunlight and 1 to 2 days if their stems are placed in water. Table XLV lists the lasting qualities of hardwood foliage with and without water.

(*c*) European hardwoods behave much the same as do their American counterparts. In fact, members of the same group of plants all over the world react in about the same way under similar conditions. It should be remembered, however, that sunlight is more intense in the tropics, and foliage will dry out and brown more quickly after cutting than will similar foliage in the United States or Europe.

(*d*) Table XLIV lists certain tropical plants, the foliage of which possesses good lasting qualities when cut.

42. FOLIAGE PRESERVATIVES.—Where wilting is a serious maintenance problem or natural growth for replacement is scarce, methods of preserving natural materials may be used.

a. Calcium chloride.—(1) Foliage treated with calcium chloride and painted with oleoresinous paint will last from 5 to 6 months. The treated and painted foliage is pliable, tough, durable, and fire retardent.

(2) The leafy branches are soaked in a solution of 6 pounds of calcium chloride to 1 gallon of water. From 4 to 6 days continuous immersion is required for foliage to become impregnated. It is then removed to dry. Shorter repeated immersions of from 2 to 24 hours reduce the total time of actual immersion required to about 30 to 40 hours. Between dippings foliage should be covered to prevent drying out, allowing further impregnation by the solution retained on the foliage. The over-all time is longer, but this method makes it possible to treat more material by rotating several

batches. Most lasting results are obtained from coarse, leathery-leafed growth such as holly, live oak, mountain laurel, magnolia, and other rubbery-leafed foliage. Finely textured leaves such as maple or birch do not attain as high a lasting quality.

(3) To test impregnation, hang some treated leaves all day to dry. If leaves are brittle, further impregnation is required to make the leaves pliable.

b. Ammonium thiocyanate and sorbitol.—(1) Complete penetration and preservation of foliage may be obtained on most deciduous leaves in from 4 to 8 hours total immersion time, and on thick leathery leaves in from 10 to 12 hours. Short—15 to 30 minutes per day—successive immersion, allowing for rotation of materials, produce similar results with less total immersion time.

(2) The mixture consists of 25 pounds of ammonium thiocyanate (commercial grade) to 10 pounds of sorbitol (commercial grade) to 100 pounds of water to 0.2 pounds of wetting agent ("Duponol," "Alkanol," or "Span" type), if available, which speeds impregnation.

(3) Solution is prepared by adding ammonium thiocyanate to the water, then stirring in sorbitol and, if available, the wetting agent.

c. Water.—The use of water will prolong the natural appearance of cut foliage (fig. 58 ① and ②). A rule of thumb to remember is that unless branches will stay green in water for at least two or three days, it is not worthwhile to supply it. Refer to tables XLIII to XLV. A quart of water a day will be ample for most branches 6 feet long. Use clear fresh water, never salt water.

43. PLANTING VINES.—*a. General.*—Vines, particularly climbing varieties, offer almost unlimited opportunities to provide cover because of their

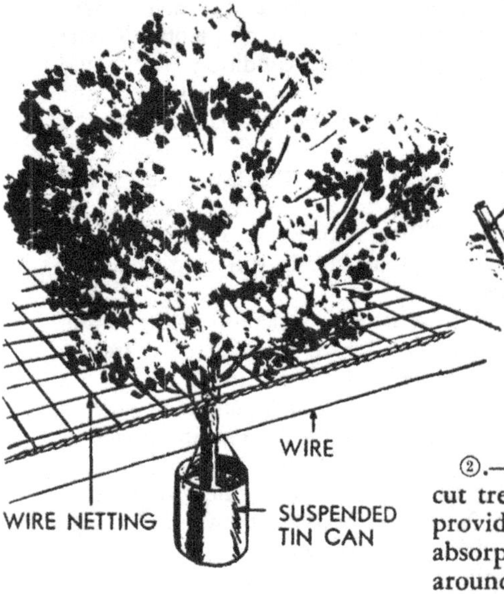

FIGURE 58 ①.—Method of supplying water to cut foliage used on overhead screens.

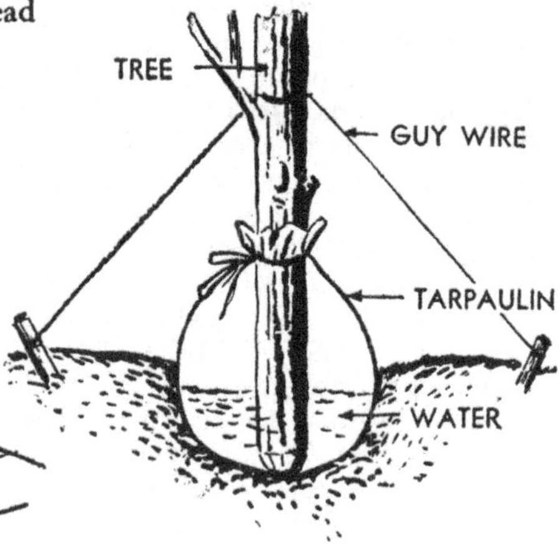

②.—Method of supplying water to cut trees. Note the tapered butt which provides greater surface for water absorption and the tarpaulin tied around trunk to prevent excessive evaporation.

TABLE XLIII.—Lasting qualities of cut foliage.

NOTE: Plants are listed alphabetically. Figures are approximate. Unless specified, values apply to both American and European species.

Plant Name	Days in Sunlight Without Water	Days in Sunlight With Water
Ailanthus (Asia)	0	0
Alder	¼	1
Apple	1	6
Ash, American	0	0
Ash, European	¼	1
Aspen	0	½
Basswood	½	4
Beech, American	½	4
Beech, European	½	6
Birch	¼	1
Black locust	0	0
Blueberry	1	6
Butternut	0	¼
Cherry, American	¼	2
Cherry, European	¼	3
Chestnut	½	3
Elderberry	0	0
Flm	¼	½
Gingko (Asia)	½	3
Hawthorn	1	6
Hemlock	4	8
Hickory	¼	1-2
Juniper	7	21
Larch, European	2	6
Larch, Japanese (Asia)	3	7
Lilac	½	6
Linden	½	4
Locust, black	0	0
Maple, American	1	2
Maple, European	1	3-6
Oak, red (group)	1	4-6
Oak, white (group)	½	4
Pine, red	14	21
Pine, Scotch	12	14
Pine, white	12	14
Plane tree	¼	2
Poplar, American	0	½
Poplar, European	¼	1
Poplar, yellow	¼	2
Privet	1	6
Red oak group	1	4-6
Spruce, Norway	14	24
Spruce, white	8	18
Sumac	0	4
Sycamore	¼	2
Thorn	1	6
Tulip tree	¼	2
Virburnum	0	1
Walnut	0	¼
White oak (group)	½	4
Willow	0	¼
Witherod	0	1
Yellowpoplar (tulip tree)	¼	2

FIGURE 59.—Native vines are used to conceal the light-colored concrete construction at this coastal installation.

easy availability, rapid growth and spread, and the comparatively little care they require (fig. 59). Table XLVI lists some of the rapid, tall-growing varieties suitable for camouflage purposes in temperate climate. Many others are useful in the tropics.

b. Transplanting.—(1) Seasons and methods of transplanting are the same as for shrubs; see paragraph 44.

c. Training vines.—Train climbing vines from earliest growth and supervise throughout the life of the plant to achieve maximum effectiveness of covering. Tie branches of vines, which have no natural method of attachment, to a support with soft twine or with a loose loop of leather or cloth. Do not use staples.

d. Propagation of vines.—The easiest and quickest methods of propagation for the majority of species are:

(1) Transplating vines, described above.

TABLE XLIV.—Lasting qualities of cut tropical foliage.

Plant Name	Days in Sunlight Without Water	Days in Sunlight With Water
Australian pine (Casuarina)	1–3	8–16
Coconut palm—leaves	3–6	4–8
Fishtail palm (Caryota)—leaves	3–6	4–8
Mangrove	½–1	½–1
Palo maria (Calophyllum)	3–6	8–16
Screw pine (Pandanus)	3–6	3–6

TABLE XLV.—Lasting qualities of cut hardwood foliage.

Hours in Sunlight Without Water	Northeastern American Plants	European Plants
8 or more	Red oak group Maple Apple Blueberry Hawthorn	Maple Apple Hawthorn Privet
5–7	White oak group Beech Basswood (linden) Chestnut Lilac	White oak group Beech Linden Lilac
3–4	Birch Cherry Elm Alder Hickory Yellowpoplar (tulip tree) Sycamore	Birch Cherry Elm Alder Ash Poplar Plane (sycamore)
0–2	Willow Black locust Ash Aspen (poplar) Butternut Walnut Sumac Elderberry	Willow Black locust Elderberry

Days in Sunlight With Water	Northeastern American Plants	European Plants
6 or more	Apple Hawthorn Lilac Blueberry	Apple Beech Hawthorn Lilac Privet
3–5	Red oak group White oak group Beech Chestnut Basswood (linden) Sumac	White oak group Maple Linden Cherry
1–2	Maple Birch Cherry Hickory Alder Yellowpoplar (tulip tree) Sycamore	Birch Plane (sycamore) Poplar Ash Alder
Less than 1	Elm Willow Black locust Elderberry Aspen (poplar) Ash Butternut Walnut	Elm Willow Black locust Elderberry

TABLE XLVI.—Vines suitable for camouflage in temperate climate.

Common and Botanical Names	Uses	Culture	Propagation	Height of Growth
Actinidia (w)	Covering flat-tops, wire fences; dense growth.	Prefers coolness, moisture, sun, or partial shade.	Seeding or layering.	20 to 30 feet.
Dutchman's Pipe (w) *Aristolochia*	Covers and hides or screens. Large foliage.	Grown in any soil with little care. Sun or shade.	Layering; cuttings.	30 feet or more, 4 to 6 feet annually.
Cinnamon Vine, Yam, or Chinese Potato (tu) *Dioscorea batatas*	Spreads rapidly over any support; dense foliage.	Easy; no care except to provide support.	Tubers or seeds.	10 to 30 feet.
Moonflower (a) *Calonyction aculeatum*	Ideal where temporary screen or covering required rapidly.	Plenty of water and sunshine.	Seeding (must be done carefully).	30 to 40 feet in a single season.
Hop (a) *Humulus*	Dense growth of foliage; must have elastic support or will pull itself out of ground.	Thrives in heat or drought; requires good soil, some care.	Seed in fall or early spring; takes 8 to 10 days to germinate.	20 feet in 3 or 4 weeks.
Honeysuckle (w, e) *Lonicera*	Dense growth where desired; cover steep banks.	Any soil.	Layering; cuttings.	From ground cover to 30 feet tall; 18 to 20 feet annually after becoming well established.
Fleeceflower *Polygonum*	Cover walls of buildings, framework; must be trained on wires.	Any soil.	Plants may be divided; root branches in soil; sow seeds as soon as ripe.	Attains 25 feet in short time.

Kudzu Vine *Pueraria*	Wherever quick screen is desired.	Prefers heavy soil; withstands drought.	Division of tuberous roots; layering if soil is kept moist.	Often attains 50 feet in a single season after becoming well-established.
Bittersweet (w) *Celastrus*	Cover walls, banks, rocky ground; tied to a stake, it may be grown to form a tree.	Very little care; almost any soil.	Cuttings of mature soft wood; layering; seeding.	3 to 10 feet.

(a) Grown from seed each year; these will die after the freezing weather of late fall.

(e) Evergreen; the leaves are not shed in winter.

(tu) Tuberous roots.

(w) Does not die to ground.

FIGURE 60.—Camouflage tree planting on a revetment for aircraft.

(2) Sowing seeds of rapid-growing annuals in the spring. Be sure that hard seed coat is cracked by filing or cutting to allow the entrance of water, thus insuring prompt germination.

(3) Root cuttings, several inches long and buried a half inch deep in boxes of sand through the winter, are taken out and planted in the spring. This method is limited to a few species.

(4) Layering is a method of propagating vines by rooting their branches. Lay the branch of a plant on the ground, cover it with soil and water it plentifully. Roots will start to grow in several places on this branch. The new plants can be severed from the old and transplanted as soon as the roots are well formed.

44. TRANSPLANTING SMALL TREES AND SHRUBS.—Transplanting involves moving trees or shrubs from one location to another in such a manner that they will continue to grow (fig. 60). Considerable care must be taken to secure satisfactory results. Among preventable causes of failure are: root systems shortened too much or otherwise mechanically injured when trees are dug up; drying out at some time between digging and transplanting; freezing of roots while out of the ground; and, after transplanting, loss of moisture from tops faster than the roots can replace it.

a. Planting seasons for transplanting trees, shrubs, and vines.—Off-season planting is to be avoided wherever possible. With great care and much added expense plants can be successfully moved at seasons ordinarily regarded as unfavorable; but will remain in a very depleted condition giving little or no leaf effect until the following natural spring growing season. Thus, frequently, nothing is gained. The most favorable times for planting are as follows:

(1) *Temperate zones.*—The best seasons for planting deciduous material are in the fall after the first frost and before freezing of the ground, and in the spring after thawing of the ground and before the appearance of

leaves. For evergreen material in more northerly zones, the most favorable seasons are in late summer after new growth has hardened and somewhat later in the spring than for deciduous plants. Farther south in the temperate zone, where the ground freezes but little, deciduous plants may be moved whenever they are leafless: evergreens may be removed during this same period.

(2) *Subtropical and tropical zones.*—In these zones there exist no such definite planting seasons, and it is possible to move most plants with fair success at any time not a period of flush growth. Palms and bamboos are peculiar in that they are most easily handled in early summer. Oaks, including live oaks, are difficult to transplant, except during the winter when they are dormant. In the sub-tropics much is to be gained by planting during cooler months, unless plants are container-grown or otherwise handled with special care.

b. Bill of materials and equipment.—Following are data for transplanting shrubs and small trees up to approximately 3-inch trunk diameter as measured 1-foot above ground:

(1) Hand tools: picks, shovels, wrecking bars, spading forks, pruning shears, knives, and saws.

(2) Trucks (4 x 4 or larger).

(3) Ordinary burlap 40 inches wide or coarsely woven cotton fabrics.

(4) Strong twine or light rope.

(5) Stakes or posts, and wire rope, or 10-gage wire, for bracing or guying.

(6) Strips of durable paper or burlap for wrapping trunks.

(7) Short lengths of rubber hose to protect trunks from guys.

(8) Peat moss or humus or equivalent at planting site.

(9) Equipment for bringing water to planting site.

(10) Wax emulsion (commercial, desirable but not essential).

(11) Issue spraying equipment.

c. Procedure in digging trees from moving with exposed root.—Deciduous trees up to 15 or 20 feet tall are usually moved with bare roots if season is favorable for moving. Larger specimens can sometimes be moved in this manner.

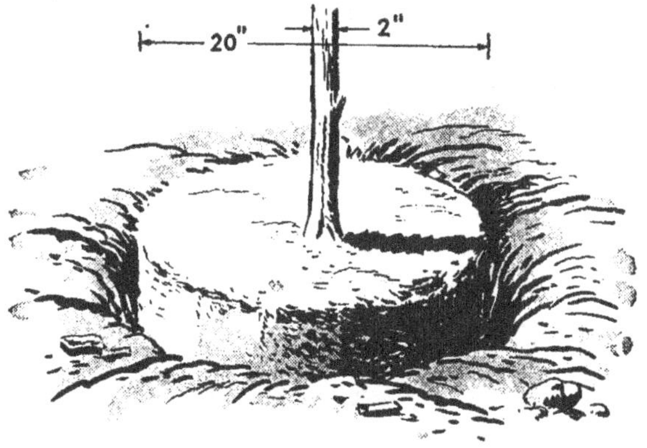

FIGURE 61.—Diagram showing method of trenching around a tree.

(1) If tree is in leaf, spray top of selected tree with wax emulsion; use care to cover underside of leaves thoroughly. This step is desirable, but not essential.

(2) Dig a trench around the selected tree outside the estimated spread of roots. A five-inch radius from the tree for every inch of trunk diameter and a depth of about 18 inches will usually avoid serious cutting of roots (fig. 61).

(3) Remove soil from roots by using a spading fork and working inward from the trench. Insert it first in the face of the trench and pull it away from the trunk with a combing action. Continue until most of the roots are exposed almost to the trunk.

(4) If the tree is exceptionally rooty and requires protracted digging, one section of the root system should be thoroughly uncovered before starting on another. Enough soil is left under the trunk to support the tree until the lateral roots have been loosened and uncovered.

(5) Lay wet burlap over the exposed roots or wrap it around them to prevent drying or other injury while digging proceeds.

(6) Tip the tree over after most of the roots have been uncovered and loosened. Use care to avoid straining or breaking roots which are still holding. If a strong taproot is encountered, dig deeper to preserve it.

d. Procedure in digging trees for moving with balled roots.—Evergreens and deciduous trees and shrubs are moved with roots kept intact in balls of soil, during unfavorable seasons, to minimize injury. The width and depth of a ball of soil depends on the formation of the roots. In most cases a width (diameter) of one foot of ball for each inch of trunk diameter is sufficient.

(1) Mark circle around selected tree a little larger than intended width of finished ball.

(2) Dig trench just outside the mark, its inner face vertical and a few inches deeper than the lower roots. When roots are encountered, they are cut off flush with face of trench using pruning shears or saw in case a clear cut cannot be made with a sharp spade.

(3) Trim off surplus soil to shape the ball. Work with back of spade

FIGURE 62.—Sketch showing burlap wrapping of a balled root system and method of fastening and binding.

toward the trunk. Trim the sides to slope inward so width of bottom is a few inches less than top. Surface of root ball is finished smooth with no projecting roots.

(4) If ball is small—less than 18 inches in width—cut under it and tip it over onto a square piece of burlap. Lift it from the hole by the corners of the burlap.

(5) Draw the burlap tightly and pin it in; lace with 8d nails. Reinforce this wrapping with strong cord laced several times around the ball and drawn up tightly (fig. 62).

(6) Balls larger than 18 inches in width are burlapped and reinforced while still in place to avoid crumbling the soil.

(7) Cut well under the ball, leaving enough soil under the center to support the tree.

(8) Place burlap around the root ball and pin it in place.

(9) Tie a loose loop of rope (top collar) around the trunk.

(10) Tie a loose loop of rope (lower collar) with a running bowline at bottom of ball.

(11) Pass the lacing cord over and under the collar ropes at intervals of 6 to 8 inches around the ball.

(12) Tighten the lower collar rope and tie it firmly.

(13) Draw up the lacing cord to strain it at each lap by taking up the slack of one lap and pulling tight on the next. This same action must be applied to alternate laps of the lace cord clear around the ball.

(14) Use a piece of strong canvas or doubled burlap to serve as a cradle for lifting and loading the ball.

(15) Do not lift balled trees by the trunk or top. Balled trees too large to manhandle are lifted and loaded by systems of skids and rollers or by special tree-moving machinery.

e. Procedure for root protection.—The sooner freshly dug trees are replanted the better, provided conditions are suitable. If delay is necessary, dormant trees can be kept for some time with little deterioration if the roots are buried or "heeled in" where the drainage is good.

(1) Make a trench as wide and deep as necessary to receive the roots.

(2) Place the trees in the trench in an inclined position. Cover roots and lower portions of the trunks with moist soil. Successive layers of trees are overlapped and covered separately.

(3) Inclined heeling-in is recommended for smaller shrubs and for trees with compact root systems. Medium trees with spreading root systems are heeled vertically. Evergreens require a shady place for heeling-in and daily sprinkling of water on their tops in warm weather.

f. Procedure in transplanting at site.—If soil is not of good quality, as determined by the native growth in the area, bring in loamy soil. Topsoil from land producing good crops may also be used. Most soils are deficient in humus and are improved by thoroughly mixing in a fourth or more by bulk of well-rotted (not fresh) manure, partially decomposed leaves, peat, or peaty woods soil.

(1) Dig holes sufficiently wide and deep to accommodate roots without cramping (fig. 63).

(2) If topsoil is to be used, pile it separately from the sub-soil. Soil used for backfill must be moist.

(3) Put in 3 or 4 inches of prepared soil on which roots will rest.

FIGURE 63.—Sketch showing relation of size of hole to tree-root system.

(4) Place a low mound of soil in the center of hole. When large trees are involved, use enough to avoid air pockets.

(5) Place tree and adjust its position so it stands at same depth as in its former location. Orient tree for best effect.

(6) Spread out roots to approximately their original position, and place soil around them to cover and hold them in place.

(7) Put in and trample successive layers of soil, being careful to prevent the tree from settling too much.

(8) When hole is nearly filled, pour in water and allow it to soak away. In humid regions, this water is not always necessary, but is recommended because it brings the soil particles in close contact with the roots.

(9) After the water has soaked down, fill the hole with loose soil. Allow the soil to settle, then fill up the hole to ground level.

g. Procedure for planting balled trees.—(1) Dig holes a foot wider and 4 to 5 inches deeper than the size of the ball.

(2) Place a low mound of earth in the center of the hole and rest the tree on it.

(3) Adjust position of tree so it stands at same depth as in its former location. Orient tree for best effect. Put in enough soil to hold it in place.

(4) Remove the ropes from the ball, pull down the burlap, and cut away the upper portion, leaving the lower portion beneath the ball.

(5) Half-fill the hole with soil and trample it to fill all air pockets.

(6) Pour in water to fill the hole. After it has soaked down, fill with soil to ground level.

(7) Make a ridge of soil 3 or 4 inches in height in an irregular ring around the trunk at the outer edge of the ball to reduce the run-off of water. Rake down this ring in winter to prevent water from standing and freezing, and causing bark injury. The ring is used for the first and second summers after transplanting.

h. Pruning.—Pruning is necessary to establish a balance between root system and top growth, especially after transplanting. If there has been a marked reduction in the number or length of roots prior to moving, there is bound to be insufficient sap for maintenance of the top. Pruning requires some skill and experience, but lacking that, tree experts may be consulted or the following suggestions used as a guide.

(1) Cut away all limbs weakened or injured by moving.

(2) Cut away enough leaf-bearing wood to balance the loss of roots in transplanting. Secondary and lateral branches are cut off close to the trunk or shortened. The leader in most cases is shortened a little. (The leader is the main stem growing vertically from the trunk. The same applies even if tree divides above trunk; the predominant limb in each division is the leader.) Preserve any short branches and twigs growing from the leader or trunk until tree becomes established.

(3) Dispose of the cut wood immediately, as it may be the breeding place of tree borers.

(4) Protect all cut and injured surfaces over 1 inch in diameter by applying commercial tree-surgery dressing or white-lead paint thinned with raw linseed oil.

i. Maintenance of transplanted trees.—(1) Trees taller than 6 or 8 feet planted in exposed areas require some support to reduce swaying by the wind, with resulting disturbance of the roots. For trees up to 15 feet tall, one or more strong stakes are driven deeply into the ground at a distance of about 3 feet from the trunk. A short length of garden hose, through which a wire has been passed, is made into a single loop to fasten the tree to the stake. Strips of webbing, strong fabrics, or soft ropes may also be used. If still stronger support is needed, use 3 stakes. Only a single loop is used for each fastening to a trunk. It must never be so tight that trunk will be constricted.

(2) Exposed trunks of newly planted trees are sometimes injured by sunscald. This is prevented by wrapping trunk and larger branches with spiral bandages made from strips of paper or burlap about 5 inches wide. The bandage is applied by starting at the top and wrapping it around the trunk spirally, overlapping the material half its width so the trunk is covered with a double thickness. This wrapping is reinforced, if necessary, by binding it with twine wound spirally in the direction opposite to that of the bandage.

(3) It is especially important that transplanted trees receive sufficient water during growing seasons for the first two years, from early spring until autumn. As a general rule, one thorough soaking each week during the growing season is sufficient. In watering transplanted trees, slowly saturate the soil until water is standing free on top of root ball. Water gently, holding hose in bucket or box so that water overflows and soaks into ground. Excess water is injurious; avoid a waterlogged condition of the soil.

(4) Apply a mulch of several inches of leaves, peat, straw, or other material to keep the surface cooler in summer and to reduce the loss of moisture by evaporation. Mulching is recommended for trees planted in the autumn where deep freezing of the soil is expected.

j. Care of transplanted trees.—Trees should be watched and inspected during the first two growing seasons after transplanting to see that they are growing as they should. Make inspections at least three times the first season and two or three times the second season. The points to be checked are:

(1) Checking new root growth by digging up with a trowel three or four small holes, evenly spaced at the outside edge of the ball, to see how much root growth has developed.

(2) Checking for air pockets by punching a number of holes down through to the bottom of the ball with a wrecking bar. Move the bar from side to side with the purpose of compacting any loosely packed soil below.

If the bar strikes a soft or easy spot, push in soil with the aid of water. Continue around the ball until settling stops. Use extreme care to avoid bruising or destroying roots.

(3) Checking soil condition as to moisture content at transplanting site with that of the original location of tree.

(4) Spreading fertilizer on top of ball and mixing in with the mulch, raking deep enough to loosen and mix about two inches of soil. Some small roots will probably be loosened also, but this will do no harm.

(5) Rebuilding ring of earth at any point where it may have been broken. Dispose of ring in freezing weather.

(6) Properly removing and disposing of any dead branches.

(7) Inspecting twigs for new growth, and foliage for density and color.

(8) Checking position of tree to see that it is not leaning. Make necessary changes of position by using guy wires.

45. TRANSPLANTING LARGE TREES.—*a. General.*—The precautions in preparation and in transplanting are the same as described in paragraph 44. In loading, moving, and unloading, additional requirements are heavy and special equipment, complete knowledge of road conditions, overhead clearances, and weights so that roads and bridges on the route may be passed safely. All these factors are investigated and the best route chosen beforehand. During loading and moving, all safety precautions possible are observed.

b. Bill of materials and equipment.—In addition to those listed in paragraph 44*b*, the following are required—

(1) Tree-mover trailer, commercial.
(2) Crane, tractor, 20-ton, 20-foot boom, issue.
(3) Three snatch blocks.
(4) Planks, 3 inches by 12 inches by 4 feet.
(5) Rollers, 3-inch steel pipe, 5 inches long.
(6) Chains and chain binders.
(7) Chain cradle.
(8) ¼-, ½-, and ¾-inch ropes.
(9) ¼-inch steel cable.
(10) ⅜- and ½-inch lag hooks.
(11) Turnbuckles (optional).
(12) Watering hose.
(13) Three-foot logs for deadmen.
(14) Six crowbars.
(15) Pole pruner.
(16) 40-yards of burlap, 40-inches wide (for each tree).

c. Procedure for digging large trees.—All large trees are moved with roots balled in burlap. Diameter of the root ball is determined by the diameter of the tree 12 inches above ground level. For general guidance, approximate dimensions of the root ball in relation to height and spread of trees are given in table LVII.

(1) Mark the outline of the ball on ground around the trunk by tying a loop of rope loosely around the tree trunk and using it as a radius.

(2) Follow method of digging and lacing ball explained in paragraph 44*d*. For convenience in working, dig trench 2 feet wide.

(3) When wrapping, leave 6- to 12-inch overlap at top and bottom and pin snugly. On very large root balls, use double wrapping.

(4) When side lacing (fig. 64 ①), leave 10 to 12 inches between

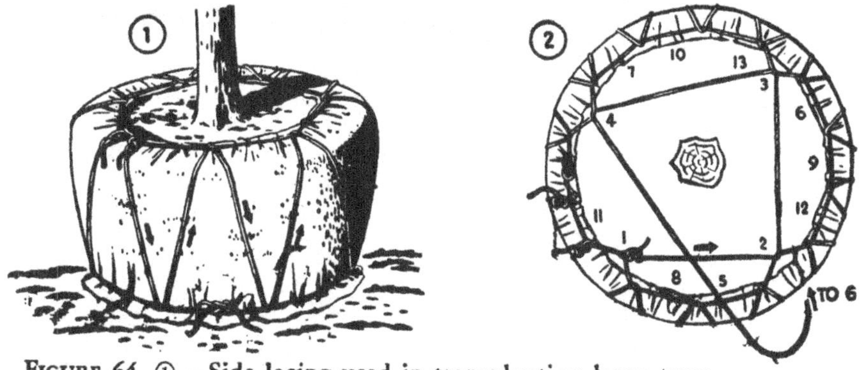

FIGURE 64. ①.—Side lacing used in transplanting large trees. ②.—Sketch of top lacing.

loops. The lower collar rope is brought up around the ball diagonally and, protected by padding, is made fast to the trunk. For additional compactness, lace across the top of the soil ball as shown in figure 64 ②, drawing rope as tight as possible.

d. Procedure in moving large trees.—The type of heavy equipment used for transporting large trees depends on the difficulties anticipated in moving, distance to be hauled, and equipment on hand. Trees to be moved only a few rods are handled with a skid or sled, rollers, and planks. Trucks and trailers are used to transport medium loads. For heavier loads, special tree-movers are required.

e. Use of skid or sled in moving medium-sized trees.—(1) To prevent cracking or crumbling of root ball, cut ball free of the ground by looping a cable around the spoil beneath the ball and winch it through.

(2) Dig a ramp on the approach side of the tree to facilitate removal from trench.

TABLE XLVII.—Relative dimensions of large trees.

Diameter of ball of earth (feet)	Approximate weight (pounds)	Diameter of trunk 12 inches above ground (inches)	Approximate height of tree (feet)	Approximate spread of tree (feet)
3	1,300	4	16	8
4	2,100	5	18	10
5	3,300	6	20	12
5½	4,400	7	22	14
6	6,000	8	24	16
6½	7,000	9	26	18
7	8,100	10	28	19
8	10,000	11	30	20
9	15,000	12	32	22
10	22,000	13	34	24
11	27,000	15	38	28
12	32,000	17	44	30
13	45,000	20	50	32

(3) Place a metal skid or improvised wooden sled in the trench in front of the root ball and at the foot of the ramp.

(4) Wrap chain cradle (which resembles a large tire chain) around top and bottom perimeter of root ball. Tighten and bind together by means of a lever-type chain binder or jack.

(5) Attach a line high up on the tree trunk, with padding. With this line, tilt tree far enough in direction opposite to ramp to permit the skid to be pushed under the root ball. Allow the tree to settle back gently on the skid, then secure it to the skid.

(6) Hook a winch cable to the skid and pull it up out of the trench and up the earth ramp.

(7) Lay down heavy planks at the end of the truck. Using steel rollers or logs between skid and planks, pull tree up onto the truck bed. Tip the tree slightly and tie in top to prevent damage to branches from overhead wires and other trees.

(8) The procedure for unloading is the reverse of loading.

f. Procedure in planting large balled trees.—Trench requirements are essentially the same as for small trees, with the following additional factors:

(1) Trenches for root balls over 3 feet in diameter are dug at least 2 feet wider than the ball.

(2) If soil does not provide natural drainage, lay 4-inch drain tile across bottom of trench leading to free drainage.

(3) For the fillback, use good topsoil free of sod, roots, or other undecayed vegetable matter. If peat moss is available, use it at the rate of 1 to 3 of topsoil or 1 to 1 of subsoil, thoroughly mixed.

(4) Fill air pocket spaces with soil; otherwise, roots will be damaged by drying out.

(5) After fill is replaced, cover the top of the ball with 2 to 3 inches of mulching material.

g. Procedure in guying large trees.—Any tree 4 to 8 inches in diameter is guyed with 3 or 4 lengths of 10-gage wire. Any tree larger than 8 inches requires heavier guy material, such as ¼-inch cable. These guys are attached to the tree halfway up the trunk either by means of lag hooks screwed into the tree or by use of rubber hose slipped over the wire to protect the bark.

(1) Insert lags into holes bored by an undersize bit. Holes are separated by at least 8 inches of height to prevent weakening the trunk at any one point.

(2) Attach guys on larger trees to deadmen sunk at least 2 feet below ground level, and away from the trunk at a distance slightly greater than the height of attachment to the tree.

(3) Use turnbuckles or racking on larger trees to keep guys taut. Set these so that about 6 inches of take-up is left.

h. Pruning, maintenance, and care of large trees are the same as already described for medium-sized trees.

CHAPTER 4

TOOLS AND EQUIPMENT

		PARAGRAPHS
Section I.	Simple Tools, Heavy Equipment, and Painting Equipment	46–55
II.	Miscellaneous Equipment	56–58
III.	Field-Expedient Manufacturing Devices.	59–60

SECTION I. SIMPLE TOOLS, HEAVY EQUIPMENT, AND PAINTING EQUIPMENT

46. GENERAL.—All types of tools from simple hand tools to heavy equipment are used in camouflage construction. Issue tools and equipment provide the mechanical assistance necessary for most work. In the absence of issue equipment many types of machines and equipment used by civilian contractors and farmers provide satisfactory substitutes. To facilitate the mass production of camouflage materials in the field, when prefabricated materials are not available, a number of field-expedient machines have been devised. Othei devices of equal value may be improvised to meet requirements of other situations.

47. SIMPLE TOOLS.—Issue tools and mechanical aids for camouflage construction are part of the organizational equipment sets of camouflage units and are listed in appropriate T/E and Engineer Supply Catalog, Part I.

48. HEAVY EQUIPMENT.—Many items of engineer issue equipment have a secondary usefulness for camouflage purposes. Because their use in camouflage is a subsidiary function, detailed data on this equipment are not covered here. Appropriate data are found in technical manuals on each item and in TM 5-255, Aviation Engineers.

49. PAINTING EQUIPMENT.—*General.*—Most paints suitable for camouflage may be applied by spraying. This paragraph does not include a list of available commercial paint-spraying equipment. The two types of spray equipment issued to troops are a portable paint-spraying unit and a mobile paint-spraying unit (pressure type).

50. PORTABLE PAINT SPRAYING UNIT.—*a.* The sprayer unit, paint, pneumatic, gasoline-engine driven, wheel-barrow-type mounting, is covered by Corps of Engineers Tentative Specification No. T-1153 (fig. 65). It is issued complete with guns, hose, accessories, tanks, tools, and spare parts which are contained in a chest. Stock number is 40-9052.5-0. Stock numbers of similar sprayers are: 40-9052.3-5; 40-9052.3; and 40-9052.7 (without compressor). This is a lightweight unit for covering small areas. It is suited for the application of oleoresinous, protein-binder, and oil-base paints, but is not normally to be used for the application of bituminous emulsion adhesive (Class A). A coverage of approximately 800 square yards in an 8-hour day may be attained. It may also be connected to a motorized air compressor.

FIGURE 65.—Portable paint-spraying unit.

b. The component parts are:

1. Gasoline-engined air compressor and air tank mounted on a wheelbarrow frame.
2. Two pressure tanks, portable, 5 gallon.
3. Two paint guns.
4. One nozzle extension for painting beyond normal reach.
5. One set of hose.
6. Accessories, spare parts and tools.
7. One chest.
8. Technical manual which covers its operation and maintenance.

c. Remarks on operation.—The paint is mixed and poured into one of the insert pails, which is then placed in the paint tank. The use of insert pails eliminates cleaning out the tanks and is convenient in changing quickly from one color or type of liquid to another. If both tanks are used, two colors of paints may be applied simultaneously. As soon as the last paint has been applied, the equipment—especially the hose and paint guns—are cleaned. This is done by filling the insert pails with a solvent of the particular paint used and blowing this liquid through the hose and guns. The guns are then taken down and cleaned. If this is not done within 15 minutes the hose becomes clogged.

51. MOBILE PAINT SPRAYING UNIT.—*a.* This portable, selfcontained unit is designed for spraying large horizontal surfaces. It is covered by Corps of Engineers Tentative Specification No. T-1291 (fig. 66). It is especially suitable for the application of bituminous emulsions, both paint and adhesive, cutback asphalt, used crankcase oil and liquids of similar viscosity. It was designed to operate directly from paint drums of 55-gallon capacity, but as most drums have bungs in the end and not on the side, the liquid must be transferred from the original drum in practice. It will operate in temperatures between minus 20 degrees F. and plus 110 degrees F. Stock number is 78-7620.5*. A coverage of 2,000 square yards in an 8-hour day has been attained.

* Not listed in Engineer Supply Catalog, Part I 1943.

b. The component parts are:

1. Gasoline engine and air compressor and 55-gallon drum mounted on a 4-wheeled dolly.
2. Two spray guns, approximately 4 feet long, with 5 interchangeable nozzles.
3. Two 20-foot hoses.
4. Attachments, spare parts, and tools.
5. Technical manual which covers operation and maintenance.

c. The equipment is not capable of being towed behind a vehicle, even at low speeds. It may be drawn by hand, without injury, but for long distances must be transported in a truck.

d. Care must be taken not to build up excessive pressure (30 pounds per square inch is normal) in the paint drums, which may have been weakened by shipment; otherwise there is danger that the drum will explode. One safety precaution is to be sure the clamps on the drum heads are tight before air pressure is built up within the drum.

52. ASPHALT DISTRIBUTORS FOR APPLICATION OF LIQUIDS OVER LARGE AREAS.—The issue distributor may be used to apply paints and adhesives over large paved or hard-surfaced areas. In the absence of this equipment, commercial asphalt distributors commonly used on road construction may be used. While equipment of this type varies in detail, it commonly consists of a tank of 800 gallons capacity to which is connected a gasoline-driven pressure pump and an adjustable pipe equipped with nozzles to deliver the material in a fan-shaped spray. A tubular heater, burning kerosene, extends through the tank, permitting the application of asphaltic material at any desired temperature. The distributor pipe is 6 feet long,

FIGURE 66.—Mobile paint-spraying unit.

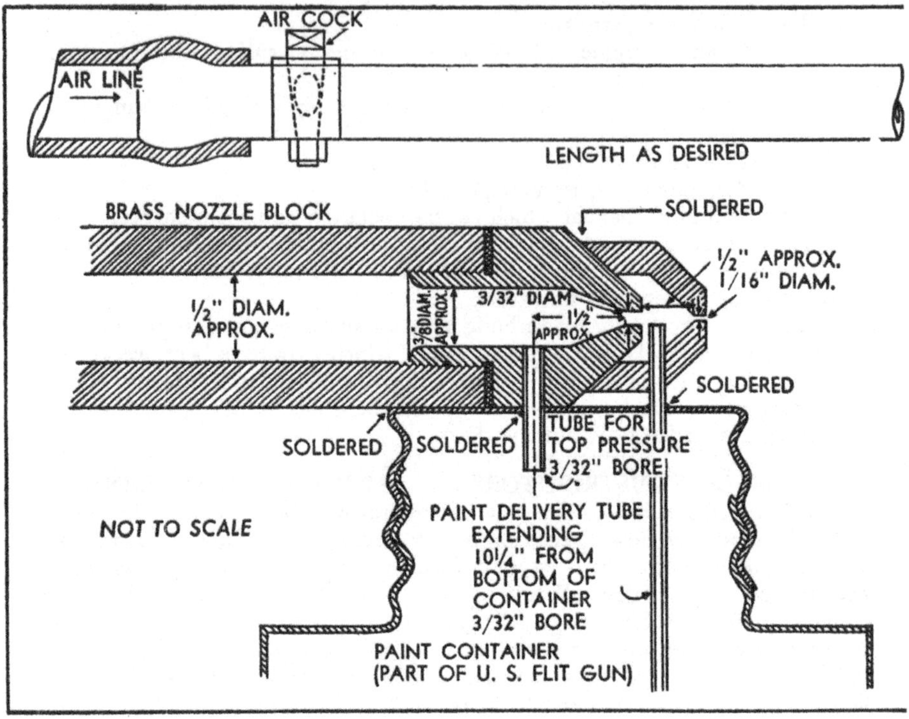
AIR COCK
AIR LINE
LENGTH AS DESIRED
BRASS NOZZLE BLOCK
SOLDERED
½" APPROX.
1/16" DIAM.
½" DIAM. APPROX.
3/8 DIAM. APPROX.
3/32" DIAM
1½" APPROX.
SOLDERED
SOLDERED
SOLDERED
TUBE FOR TOP PRESSURE 3/32" BORE
NOT TO SCALE
PAINT DELIVERY TUBE EXTENDING 10¼" FROM BOTTOM OF CONTAINER 3/32" BORE
PAINT CONTAINER (PART OF U. S. FLIT GUN)

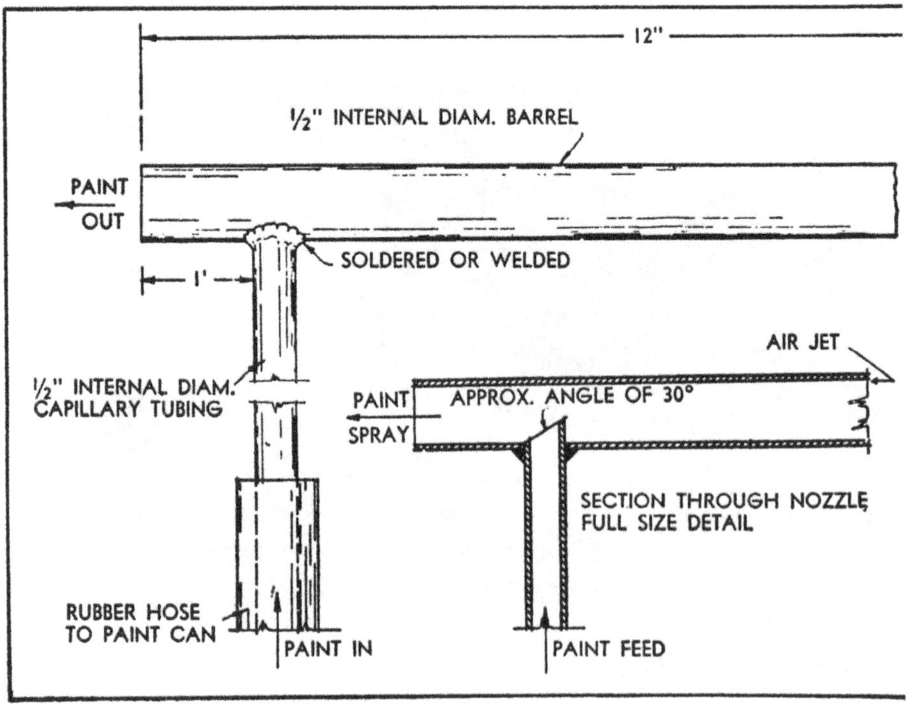
12"
½" INTERNAL DIAM. BARREL
PAINT OUT
SOLDERED OR WELDED
1'
AIR JET
½" INTERNAL DIAM. CAPILLARY TUBING
PAINT SPRAY
APPROX. ANGLE OF 30°
SECTION THROUGH NOZZLE FULL SIZE DETAIL
RUBBER HOSE TO PAINT CAN
PAINT IN
PAINT FEED

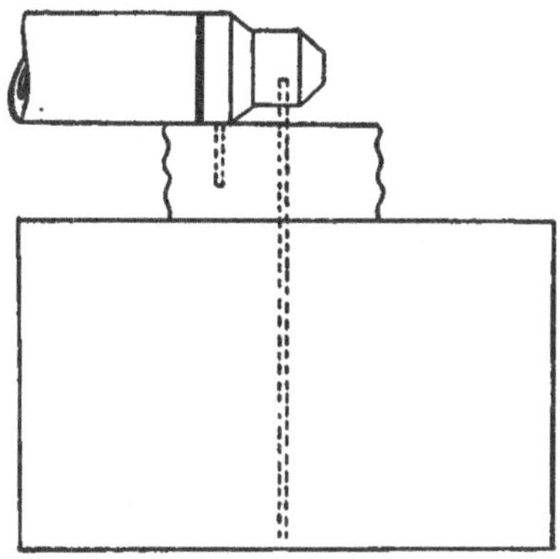

FIGURE 67 ①.—Field-expedient sprayer. Paint is drawn from container by suction.

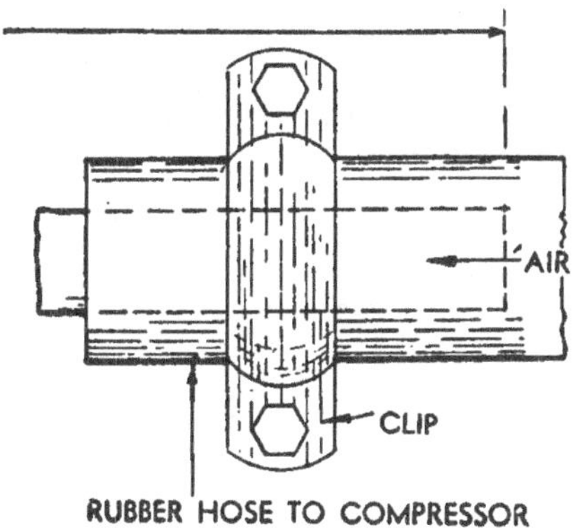

FIGURE 67 ②.—More elaborate paint sprayer, with flit-gun container for paint receptacle.

FIGURE 67 ③.—Expedient sprayer with standard type valves for the adjustment of spray. To provide a flat spray the pipe extension may be flattened as shown.

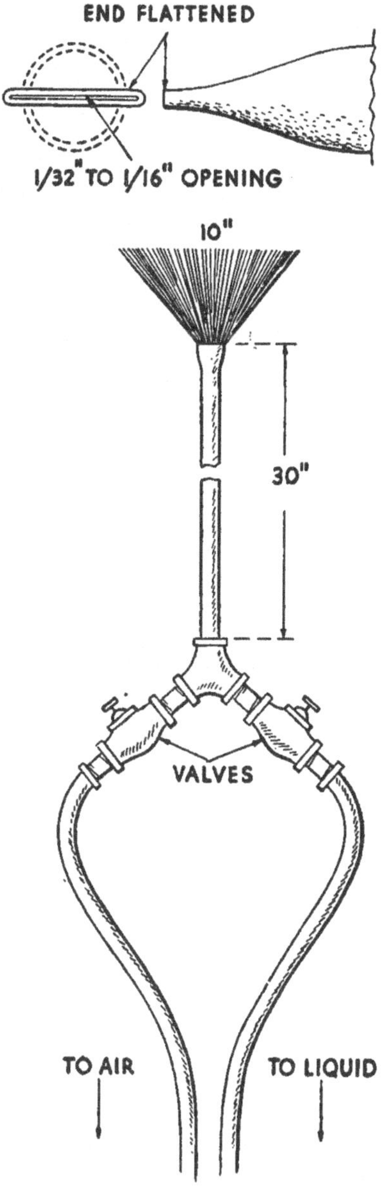

carries 12 spray nozzles, and is equipped with a 2-foot extension at each end, each extension carrying 4 spray nozzles. Hand-operated flexible extensions are provided for reaching places that cannot be reached by the distributor pipe. The whole unit is mounted on a 4-ton truck. A dial in front of the driver is calibrated in gallons per yard (or similar units) and permits the driver to distribute liquid at a uniform rate by regulating the speed of the truck.

53. COMMERCIAL EQUIPMENT.—When issue spray equipment is not available civilian and commercial sources may be able to supply it. Considerable variation in any types of commercial paint sprayers will be encountered, but when need arises almost any spraying device may prove useful. Sprayer equipment not normally intended for the application of paint, such as fire extinguishers, stirrup pumps, golf-course spray equipment, disinfectant or orchard sprayers, may be locally procurable and serve or be adapted to serve the purpose. Care must be taken to determine which equipment is adaptable to the application of the particular paint or adhesive desired.

54. FIELD-EXPEDIENT PAINT SPRAYERS.—*a.* The improvised sprayers shown in figure 67 have been successfully used in the field.

(1) Three sprayers of the type shown in figure 67 ① may be worked off one compressor truck. To operate, insert the paint intake tube, approximately 2-feet long, through a hole near the top edge of one side into the paint container, a half gas can with handle, carried by the operator. The end of the tube should reach as near the bottom of the can as possible. The spray gun is held slightly below the level of the paint container to provide some gravity feed until sufficient air suction is obtained. A mixture of 2 parts water to 1 part paint by volume is used. Dilution to this extent is necessary to enable the gun to work satisfactorily. A second application is necessary at times. Coverage averages 30 square yards per gallon. Three men, each with a spray gun, can cover 300 square yards in 1 hour.

(2) Figure 67 ② illustrates a more elaborate field-expedient sprayer developed in the field.

(3) Figure 67 ③ illustrates an alternate sprayer made from ordinary plumbers' supplies. Valve connections, pipe, and hose used are 3/8 inch to 1/2 inch.

55. PAINT MIXERS.—*a. Pneumatic paint agitator.*—An issue pneumatic agitator for mixing paints is covered by Corps of Engineers Tentative Specification T-1575. Stock number is 40-1132.7-5. The component parts are:

(1) Air motor connected to a 27-inch propeller shaft.
(2) Two propellers.
(3) A clamping device for attaching agitator to a 50-gallon drum.
(4) Spare parts.

b. Motor-driven paint agitator.—This issue apparatus is similar to the above except that it is electrically powered and is equipped with a 50-gallon mixing tank. It is covered by Corps of Engineer Tentative Specification T-1564, Stock No. 40-1132.3-7. The component parts are:

(1) Mixing tank with outlet valve.
(2) Mixer, electric, powered with 2 propellers.

SECTION II. MISCELLANEOUS EQUIPMENT

56. DISTRIBUTORS FOR TEXTURING MATERIALS.—*a.* No distributors for texturing materials are issued. Granule guns are available com-

mercially in the United States. These guns are suitable for use in distributing texturing materials such as slate granules. The sprayer consists of a metal tank mounted over a small electrically driven blower. Granules are distributed evenly and economically on the adhesive through a metal pipe about 2 feet long. The apparatus is equipped with a shoulder strap for ease in use. Some types are equipped with an open granule container which permits the replacement of granules without interrupting operation of the machine.

b. Distribution of texturing on a large scale may make use of any suitable, commercial, road-building equipment or farming machines, such as manure distributors.

c. One field-expedient texturing distributor has been adapted from an ensilage blower mounted on a frame, its outlet horizontal to the ground (fig. 68). An 8-horsepower gasoline engine, also mounted on the frame, is used as a power unit to operate the blower. A hopper for loading from a dump truck to the blower is arranged at approximately a 45° angle to facilitate the transfer of chips from the truck bed to the blower. The entire unit is mounted on wheels and has a pin-type hitch to fasten the rig to the rear of a truck. Attached to the blower discharge is a flexible 6-inch tube, 5 feet in length, which allows chips to be sprayed with a sweeping motion, the operator having full control of the rate of application. The platform on the blower unit is so arranged that the operator of the spreader can ride on the machine as it is being towed by a truck. The rate of application of wood chips by this method averages approximately 40,000 square yards per 8-hour day. Any light-weight material, such as wood chips, may be applied with this machine.

FIGURE 68.—An improvised granule distributor.

57. WOOD-CHIPPING MACHINE.—A pilot model of a wood-chipping machine for mass production of texturing granules is now being tested.

58. ROTARY CLOTH-CUTTER BLADE.—Electrically powered, hand-operated cloth cutters are issue equipment. It is a time-saving tool in the manufacture of large quantities of garnishing strips from bolted fabric or in the mass production of cloth-covered decoys. For quantity production—as in the manufacture of covering-fabric for decoys—it is advisable to cut a master pattern which will prevent wastage of material on the width supplied. This cutter will cut through approximately 20 layers of burlap. It is covered by Corps of Engineers Tentative Specification T-1566. Stock number is 40-1825.5-55. To supplement issue machines, civilian cutting machines may sometimes be found in factories of clothing manufacturers and stationers.

SECTION III. FIELD-EXPEDIENT MANUFACTURING DEVICES

59. GENERAL.—Often before raw materials can be used for camouflage purposes they require treatment. To process them in large quantities, manufacturing equipment for field use has been improvised. Similarly improvised machines to speed up the application of prefabricated materials have been developed. This section describes some of the most practical of these field-expedient devices and machines, but it must be pointed out that this list can be expanded and the devices changed to fit different situations. It is impossible in this manual to discuss the conversion of farming and civilian equipment to camouflage purposes since the types available all over the world vary considerably. Officers charged with large-scale camouflage installations will, however, be well advised to investigate the possibility of converting such equipment to related but different uses. For example, silage machines mav be a means of obtaining texturing materials, rock crushers and primitive flour mills may be adapted to producing dry pigment, and seeders and manure spreaders adapted for distributing texturing materials.

FIGURE 69.—Garnishing wire netting in factory operated by engineers in zone of the interior.

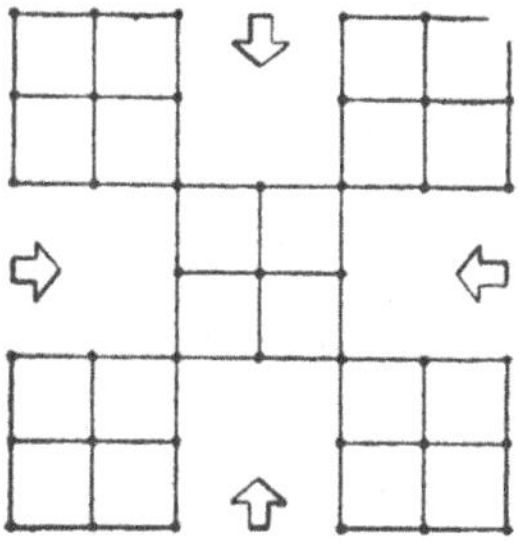

FIGURE 70 ①.—Diagram of table for garnishing twine nets

②.—Civilian employees garnishing twine nets in factory operated by engineers.

60. CAMOUFLAGE FACTORY METHODS IN THE FIELD.—*a.* Camouflage materials are often produced extensively in rear areas to supply the needs of troops in forward areas. This section describes methods which have been employed for various common operations. The information is basic; variations and improvements will be made depending on the situation and construction material available at a particular site.

b. Garnishing wire netting—factory method.—(1) Construct a framework resembling a table (fig. 69). The width should be about 1½ feet more than the widest netting that will be garnished. This allows strips of garnishing to be hung at the ends of the sectional frames, within easy reach of personnel. Height should allow personnel to work at the frame comfortably.

(2) At each end of the frame table is a supporting rack for a pipe or steel rod on which a roll of netting is unwound and wound.

(3) A roll of wire netting is placed in one of the supporting racks, unrolled over the framework table, garnished, then re-wound in the rack at the opposite end.

FIGURE 70 ③.—Troops garnishing twine nets.

④.—Troops spray-painting newly garnished nets.

(4) The netting is garnished solidly; thinning of garnishing is done by using troops.

c. Garnishing twine nets—factory methods.—(1) Construct a framework table, 15 by 15 feet square, with four bays (fig. 70 ① and ②). Larger tables are not recommended because of the great amount of lumber and space required. Larger nets can be garnished by moving them about on the table as required.

(*a*) Place twine net on the table. Garnish from the center outward, thinning according to specifications. This is accomplished by entering the bays and moving out of them as garnishing progresses toward the sides.

(*b*) Place supplies of colored garnishing strips within easy reach of personnel.

(*c*) Garnished nets can be folded at the tables or removed to other areas for folding.

(2) Another method for mass production employs a garnishing frame approximately 48 feet long and about shoulder-high (fig. 70 ③). Fourteen men can work conveniently on the frame—seven to a side—under the supervision of NCO's.

(3) It may be estimated that with troop labor approximately 200 square feet of garnishing will be produced per man in an 8-hour day. Over an extended period, an average of 25.6 square feet of garnishing per man-hour has been realized. The maximum production for a single day was 26.6 square feet per man-hour. Approximately 1,200 garnishing strips are required to garnish 1,000 square feet of net with 100% coverage.

(4) Spray-painting garnished nets is occasionally necessary. The apparatus illustrated in figure 70 ④—which uses a 35 foot boom—has been successfully used. A detail of three men is used, one on the motor and two alternating on the spray gun. Goggles and respirators should be worn because fine paint vapor is injurious to lungs and eyes.

d. Painting and cutting bolted fabric for garnishing—factory method.— (1) When it is necessary to process large quantities of bolted fabric to make garnishing strips, a field factory is set up. Figure 71 ① illustrates the organization of such a factory. Bales of uncolored fabric are brought in by truck and piled at **A** within convenient distance of the paint vats **F**. The supply of paint is spotted at **B** near the paint mixers **C** and the water supply **D** which should, if possible, be piped in. Bolts of fabric are laid out at **E** and the mixed paint carried by bucket to the paint vats **F**. The fabric is then passed into the paint vat, through the 44-inch wringer and wound on the reel **G**. Here it is cut into panels by electric cutters and carried to **H** near the cutting tables **J**. The panels are aligned, preparatory to cutting, by being suspended from a series of nails driven through the board at **I**. The panels are then removed, folded once and placed on the cutting table where they are cut into 2-inch widths by electric cutters. The cut strips are placed in the racks **K** and tied into bundles when a rack is filled. The bundles are then carried to **L** if the net-garnishing operation is at another site, or directly to the net-garnishing frames if these are nearby.

(2) With the addition of a simple stand located in front of the paint vat, the same procedure may be used to recolor 2-inch width rolls. A length of pipe inserted through the holes in the rolls and supported by the stand acts as an axle as the strips are run through the vat. Fifteen rolls may be processed at a time.

(3) The construction and operation of the special equipment which must be built by the unit setting up the factory is shown in figure 71 ②, ③ and ④.

(4) The time lag in the process is usually sufficient to allow garlands to dry before the nets are completely garnished. If, however, it is necessary to store garlands before they are used as garnishing, they must be dried, because damp fabric rots quickly. If this procedure is used to color full-

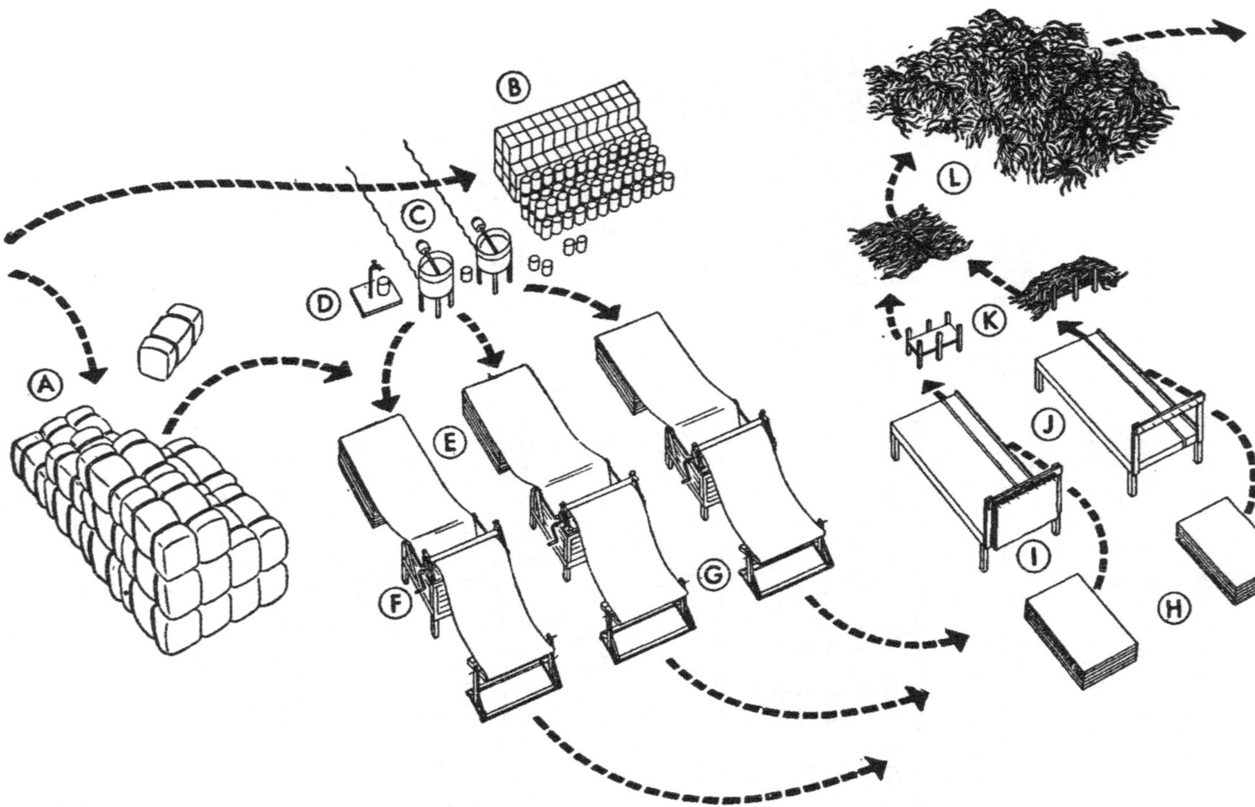

FIGURE 71 ①.—Operation diagram of field shop for coloring fabric and making garnishing strips.

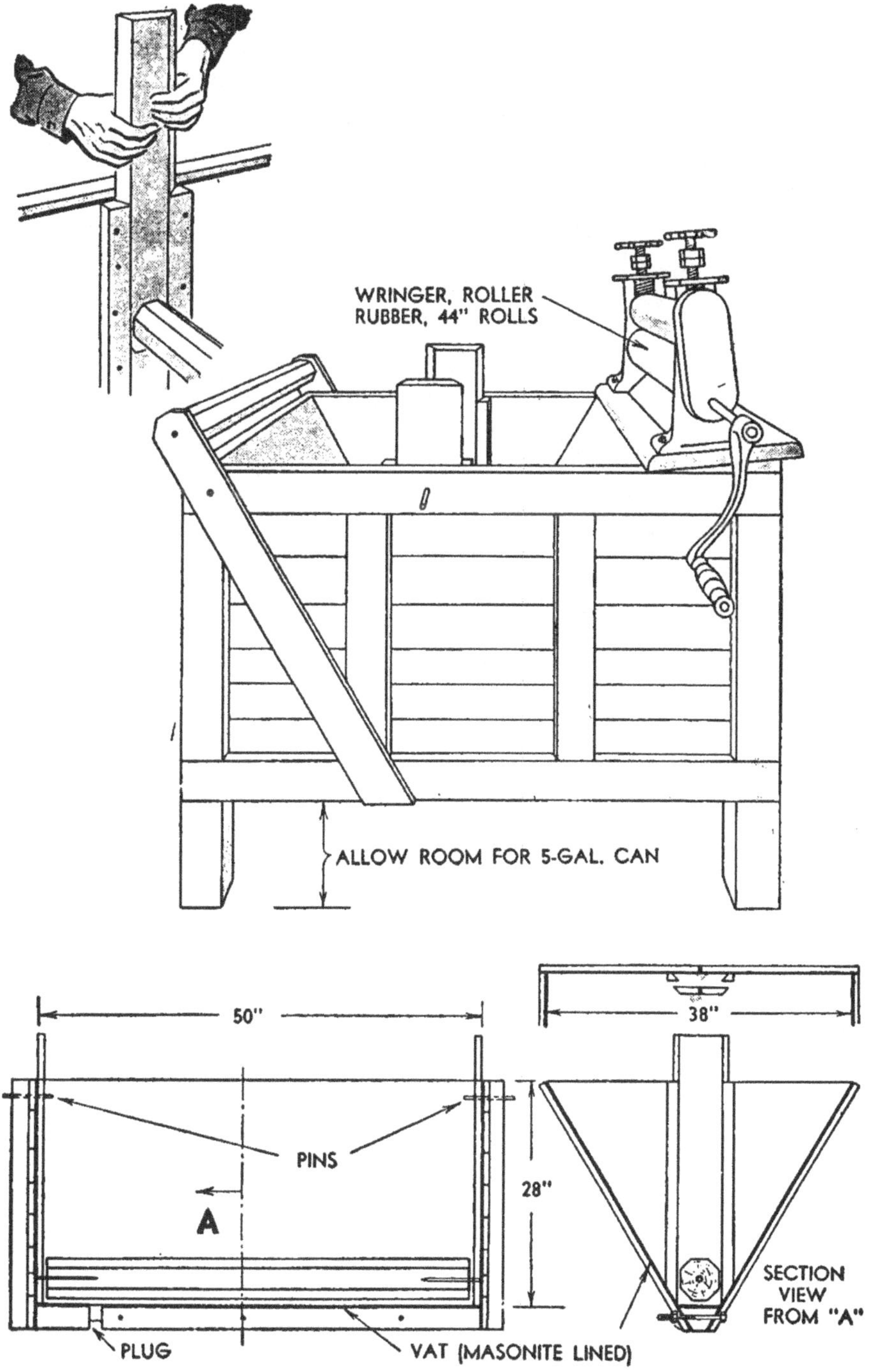

②.—Construction diagram of paint vat.

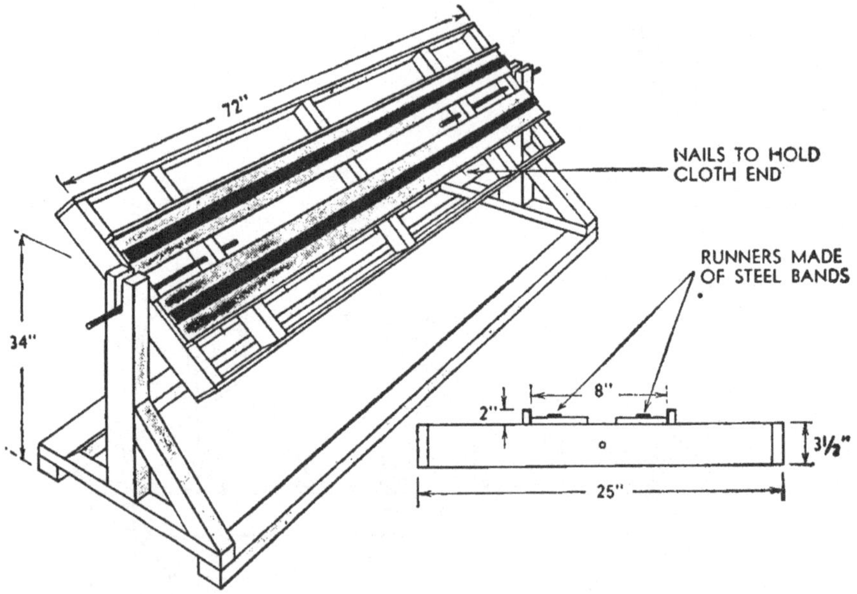

FIGURE 71 ③.--Construction diagram of reel.

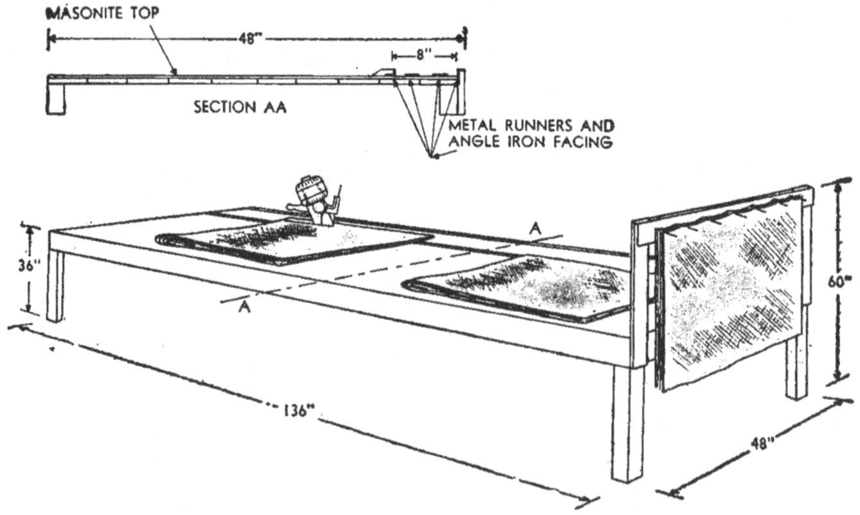

④.—Construction diagram of cutting table.

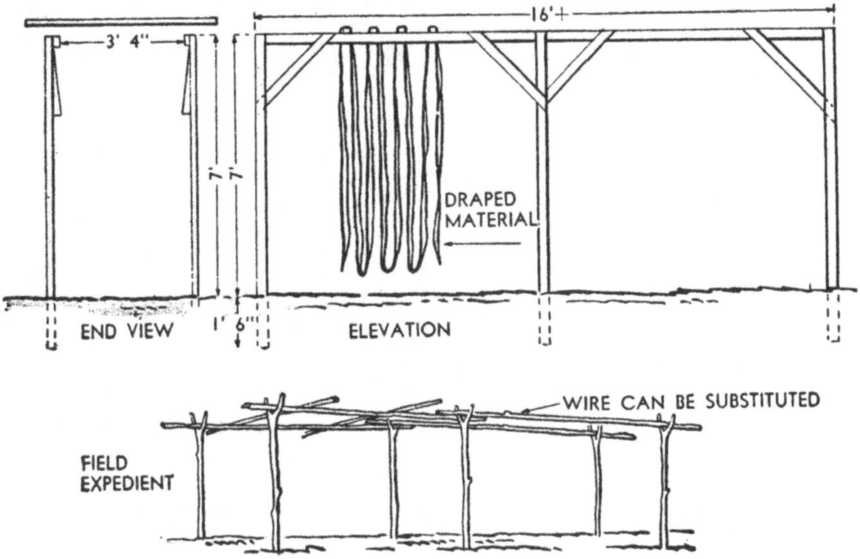

FIGURE 71 ⑤.—Construction diagram of drying rack.

width bolted fabric for screening purposes, the use of a drying rack, for the same reason, is recommended (fig. 71 ⑤). It is to be noted that approximately 1/3 greater volume of production is realized when fabric is cut while still damp.

(5) Tools and equipment include:

3 Paint vats, 25-gallon capacity, built by unit (fig. 71 ②).
3 Wringers, roller, rubber, 44-inch.
3 Reels for winding dipped fabric, built by unit (fig. 71 ③).
5 Electric cloth cutters, (three extra cutters are needed to supplement the two presently authorized a Camouflage Battalion (Army)).
2 Cutting tables, built by unit (fig. 71 ④).
2 Machines, paint-mixing, 50-gallon capacity.
1 Generator set, portable 3 KV-A.
— Water supply—piped in, if possible.
3 Stands for rolls of garnishing strips (if required).
1 Drying rack (length determined by circumstances) (optional).
— Hatchets, shears, knives (if electric cutters are unavailable).
2 Racks for garlands, built by unit (Fig. 71 ①, K).

(6) Cloth painting and cutting detail may be organized as follows:

*Enlisted Men**	*Duty*
3	Mix paint.
1	Guide bolted fabric into paint vat.
1	Guide painted fabric into wringer.
3	Operate wringer (alternating).
2	Wind cloth on reel.
1	Cut reeled cloth.
2	Bale and carry cut cloth.
5	Per electric cutter at cutting tables to cut garnishing strips.
2	NCO supervisors.

* The total number of men may be increased to 47, if circumstances warrant.

FIGURE 72 ①.—Troop clothing placed in net.

②.—Dipping clothing in dye vat.

③.—Rinsing dyed clothing with fresh water.

(7) The chart which follows gives an approximate basis for estimating rate of production. With this procedure 112,500 garlands have been manufactured per 8-hour day.

Process	*Unit of Production*	*Man-hours per Unit*	*Number of Men*
Painting 40-in. bolted burlap (including mixing paint)	100 yards	2.50	12
Cutting garlands from 40-in. bolted burlap			
(a) Hatchet method	900 garlands	8.00	5
(b) Electric cutter	900 garlands	1.50	5
Cutting 2-in. rolls with electric cutter	900 garlands	0.70 to 1.30*	6

* Variation is caused by rolls of several colors; some of which are easier to cut than others.

e. Dyeing troop clothing—factory method.—(1) Tie knots in clothing so that dyeing will result in a mottled rather than a solid color.

(2) Place clothing in a large twine net (fig. 72 ①) and lash the four corners together at the top of the bundle.

(3) Procure steel vats; raise them on rocks or masonry high enough to allow for fire pits under them.

(4) Prepare a dye solution (table XVII) of the required color and pour it into the vats.

(5) Build fires under the vats and heat solution to the proper temperature.

(6) Run a pipe, steel rod, or strong wooden pole under the lashing at the top of each bundle before immersing in vat. These rods or poles rest on the sides of the vat, thus supporting the load and preventing the bundles from being burned by touching the bottom of the vat.

(7) Use a vehicular-mounted boom crane, or construct a boom derrick, for hoisting the bundles of clothing and lowering them into the dye vats (fig. 72 ②).

(8) Stir the dye solution constantly as long as the clothes are immersed.

(9) Remove bundles by crane or derrick.

(10) Rinse thoroughly (fig. 72 ③).

f. Making garnishing strips—factory method.—(1) Construct equipment as illustrated in figure 73.

(2) Place roll of burlap on feed roller. Pass free end of burlap under guide roller, through the row of razors, then tack down each strip to the

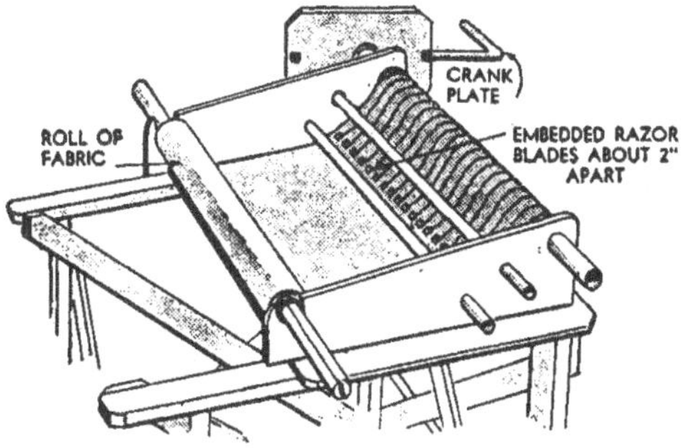

FIGURE 73.—Machine for cutting bolted fabric into 2-inch garnishing strips.

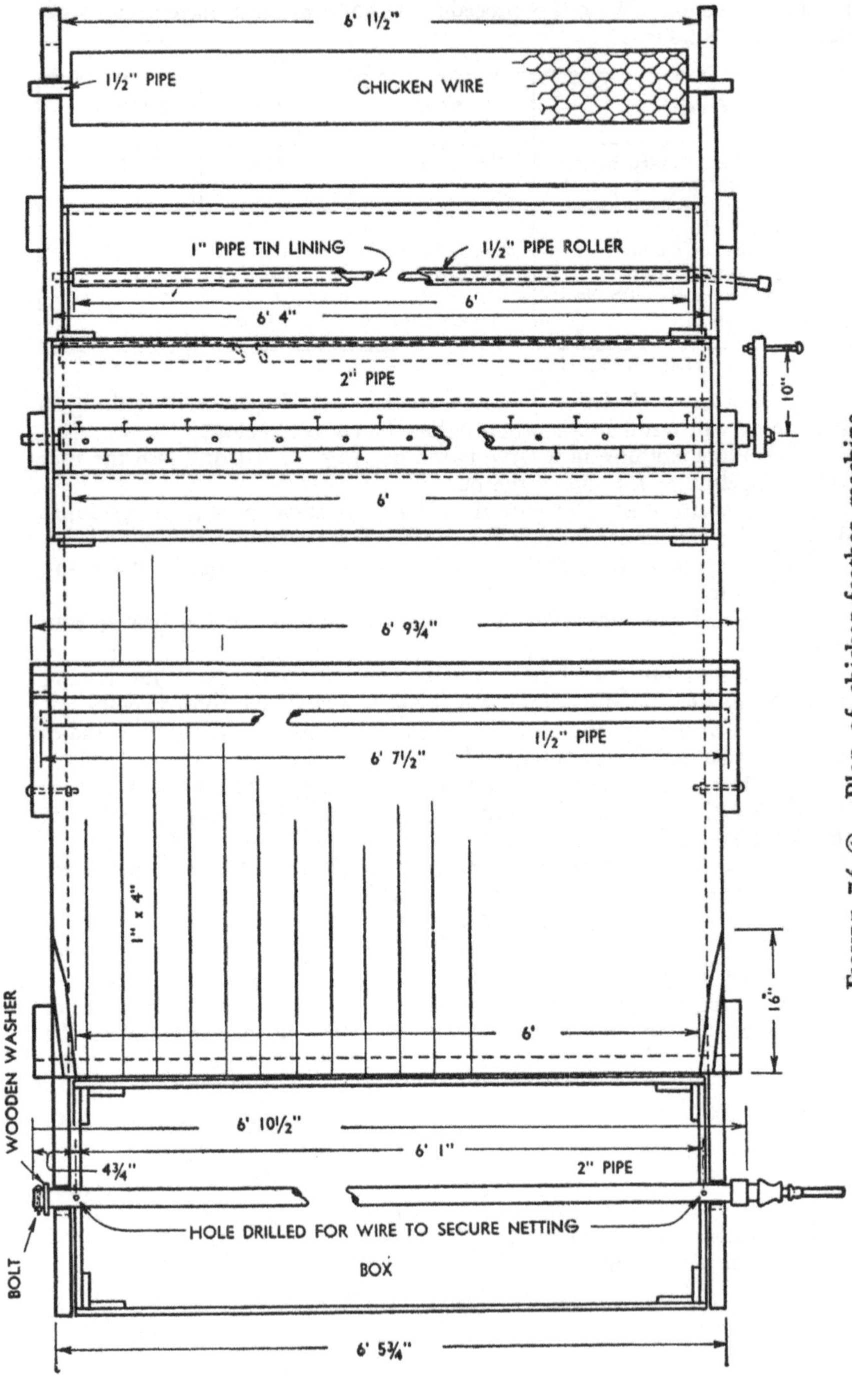

FIGURE 74 ①.—Plan of chicken-feather machine.

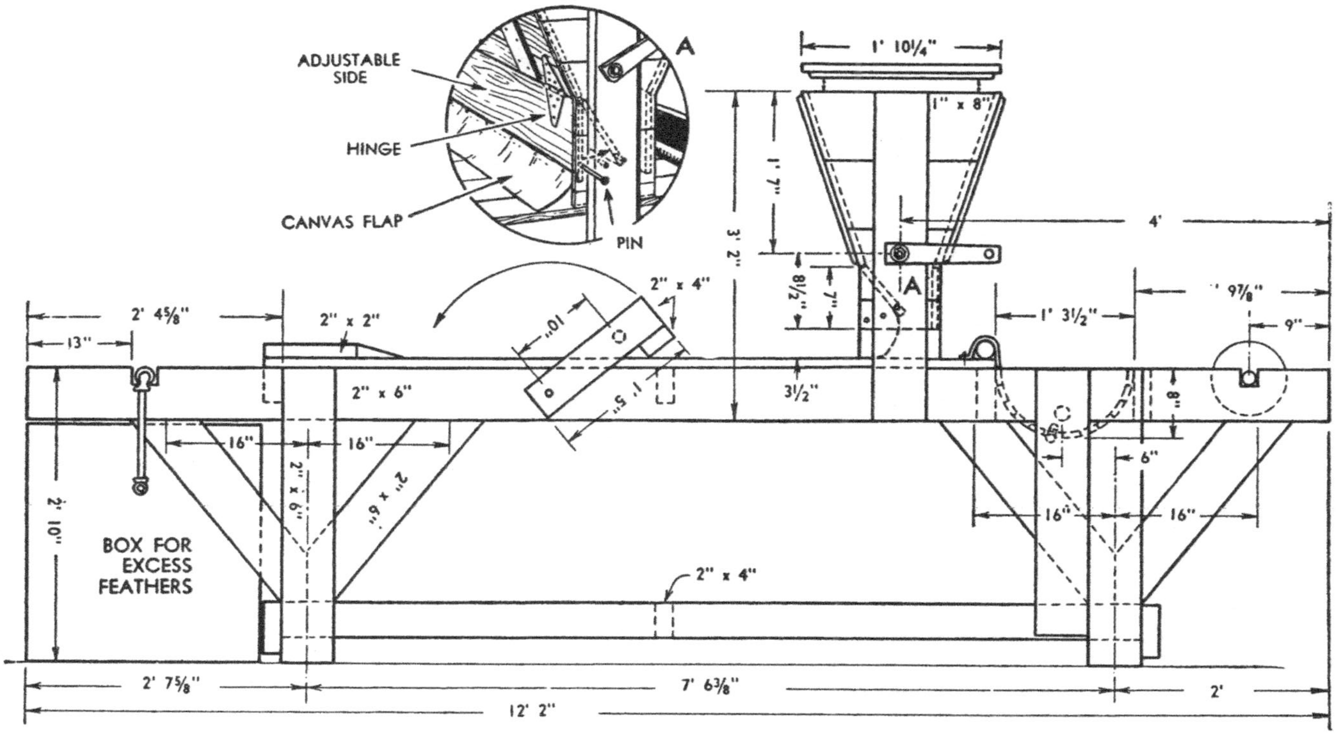

②.—Elevation of chicken-feather machine. A is a detail of feather hopper showing adjustable side regulated by pin which controls flow of feathers by increasing or decreasing size of opening in hopper-throat. Canvas flap keeps feathers from blowing back.

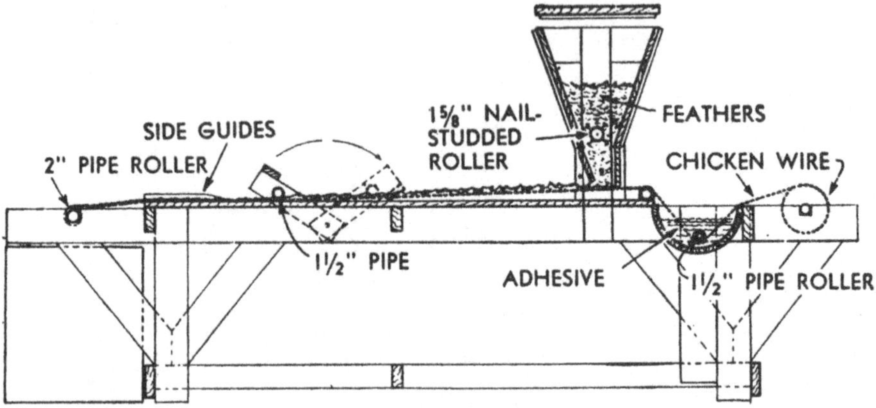

FIGURE 74 ③.—Chicken-feather machine in operation.

wind-up roller and crank the remainder of the burlap roll past the razor blades.

g. Applying chicken feathers to wire net—factory method.—Construction of a feather-garnishing machine is shown in figure 74 ①, ②, ③. In operation, it requires about five man-hours per 100 linear feet of wire netting to affix and color the feathers. The bill of materials is given in table XLVIII. The operation of the machine is covered in paragraph 23*d* (4) (*a*).

h. Other field expedients for manufacture of camouflage material.—

(1) Tin-can cutter for tin-can garnishing (fig. 75).

(2) Portable-screen field loom, which is constructed as in figure 13. Operation of the loom is described in paragraph 16.

(3) One method of pulverizing colored rocks for manufacturing dry

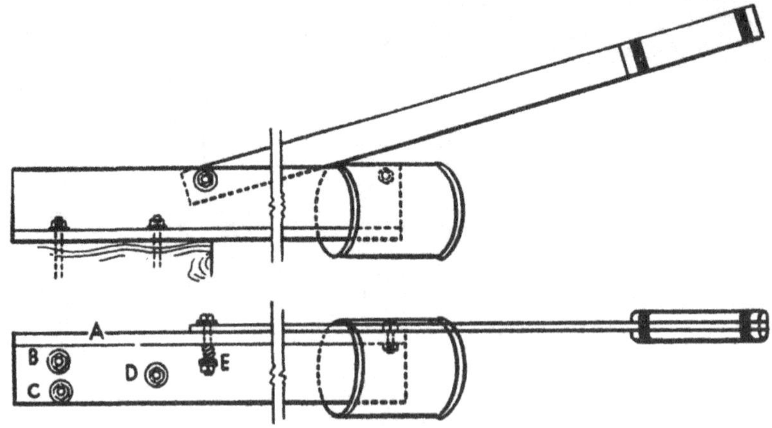

FIGURE 75.—Tin-can cutter. Material used for construction includes: 3-inch angle iron, one piece of steel, and a few bolts. Angle iron A is fastened to wood block on tree stump at three points B, C, D. Blade is made of spring steel and fastened to angle iron by bolt and spring E. In operating cutter, slip No. 10 can over end of angle iron as shown and cut into strips, holding and rotating can as each cut is made.

TABLE XLVIII.—Feather machine—bill of materials.

Material	Size	Length	Number	Use
Lumber	2" x 6"	12' 2"	2	Top side frames.
	2" x 6"	2' 10"	4	Legs.
	2" x 6"	2' 8"	8	Diagonal braces.
	2" x 6"	3' 2"	2	Hopper posts.
	2" x 6"	6' 1½"	2	Top frame spreaders.
	2" x 6"	6' 5¾"	2	Bottom frame spreaders.
	2" x 6"	2' 6"	1	Tank, drain support.
	2" x 4"	6' 1½"	3	Bridging.
	2" x 4"	8' 6½"	2	Bottom side frames.
	2" x 4"	6' 9¾"	1	Compactor roller spreader bar.
	2" x 4"	1' 5"	2	Compactor roller sides.
	2" x 2"	1' 4"	2	Guides
	2" x 2"	7'	1	Feather agitator bar.
	1" x 8"	6' 5¾"	11	Hopper side and top.
	1" x 8"	1'10¼"	2	Hopper ends.
	1" x 8"	1' 7"	2	Hopper ends.
	1" x 8"	1' 1"	2	Hopper ends.
	1" x 8"	0' 9½"	2	Hopper ends.
	1" x 8"	1' 3½"	2	Tank sides.
	1" x 4"	6' 9½"	20	Table top.
	1" x 4"	1' 9"	2	Hopper top battens.
	1" x 4"	1' 8"	4	Hopper side battens.
	1" x 2"	6' 2½"	12	Adhesive tank.
	1" x 2"	1' 1"	1	Hopper crank arm.
Pipe	2"	2" to 1"	1	Reducer.
	2"	Short	1	Nipple.
	2"		1	Coupling.
	2"	6'10½"	1	Rerolling bar.
	2"	6' 4"	1	Tank guide bar.
	1½"	6' 7½"	1	Compactor roller bar.
	1½"	6' 8"	1	Netting roll axle.
	1½"	6' 0"	1	Tank roller.
	1"	90°	2	Elbows.
	1"	....	1	Drain pipe cap.
	1"	6' 4"	1	Tank roller axle.
	1"	1' 0"	1	Crank.
	1"	8"	1	Crank handle.
	1"	6"	1	Drain pipe.
Bolts and Nuts	3" x ⅜"		1	
	7" x ⅞"		1	
	5" x ⅜"		2	
Sheet Galvanized Iron	24 gage	84" x 36"	1	Adhesive tank.
Nails	8d		5 lbs	
Hinges			3	
Canvas		72" x 6"	1	
Box			1	Container for excess feathers

pigment employs a mortar and a pestle (fig. 76). To make a mortar, weld a 2-foot length of 6-inch iron pipe to a flat iron plate. The pestle is made of a 4-foot length of 1½-inch pipe with a pipe cap on one end. Pigmented rock is placed in the mortar and pounded with the pestle.

(4) Another method of manufacturing pigment is to:

(*a*) Drill a hole through a 55-gallon oil drum, insert in the hole

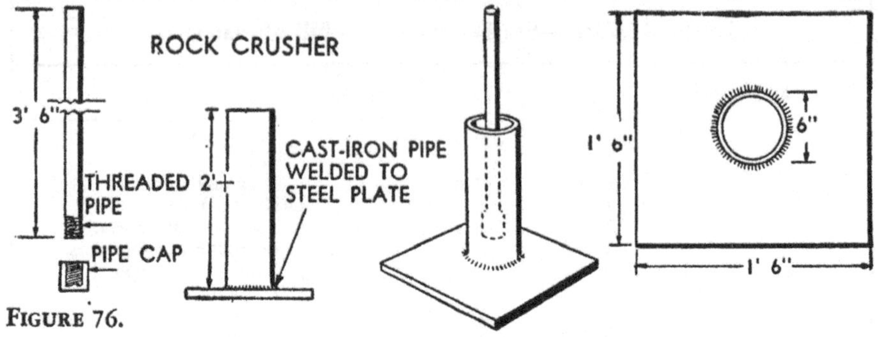

FIGURE 76.

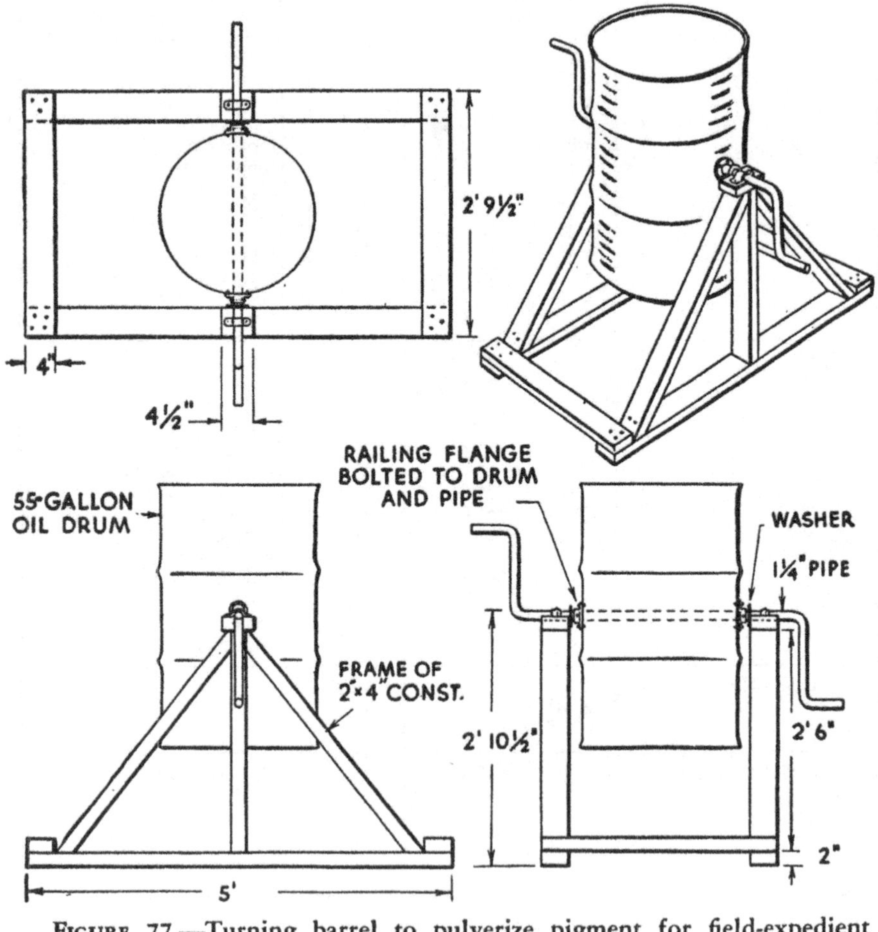

FIGURE 77.—Turning barrel to pulverize pigment for field-expedient paint.

a length of pipe to serve as an axle, and fasten it as shown in figure 77.

(*b*) Attach handles to ends of the pipe so that the drum can be revolved. Place the drum on a frame so that it is supported by the axle.

(*c*) Place ten round granite rocks, each about 4 inches in diameter, inside the drum. Add two bucketsful of roughly broken pigmented material.

(*d*) Close the top of the drum and turn the drum until the pigment has been well powdered.

(5) A field-expedient pigment-collecting chamber (fig. 78) may be constructed as follows:

(*a*) A box-like structure, 72 inches high with a 19- by 19-inch section, is made with one side removable. Crushed pigment is poured into the top end. At the bottom is a drawer to catch heavy pieces of pigment. The light pieces settle on metal collecting shelves, 2½-inches deep, which are fastened to the inside of the chamber on 9-inch centers.

(*b*) For best results the chamber should be plumb. The heavy material which falls into the drawer at the bottom are re-poured through the top until all the light particles have been collected in the collecting shelves. Then the removable side of the chamber is opened and the usable pigment removed from the collecting shelves.

FIGURE 78.

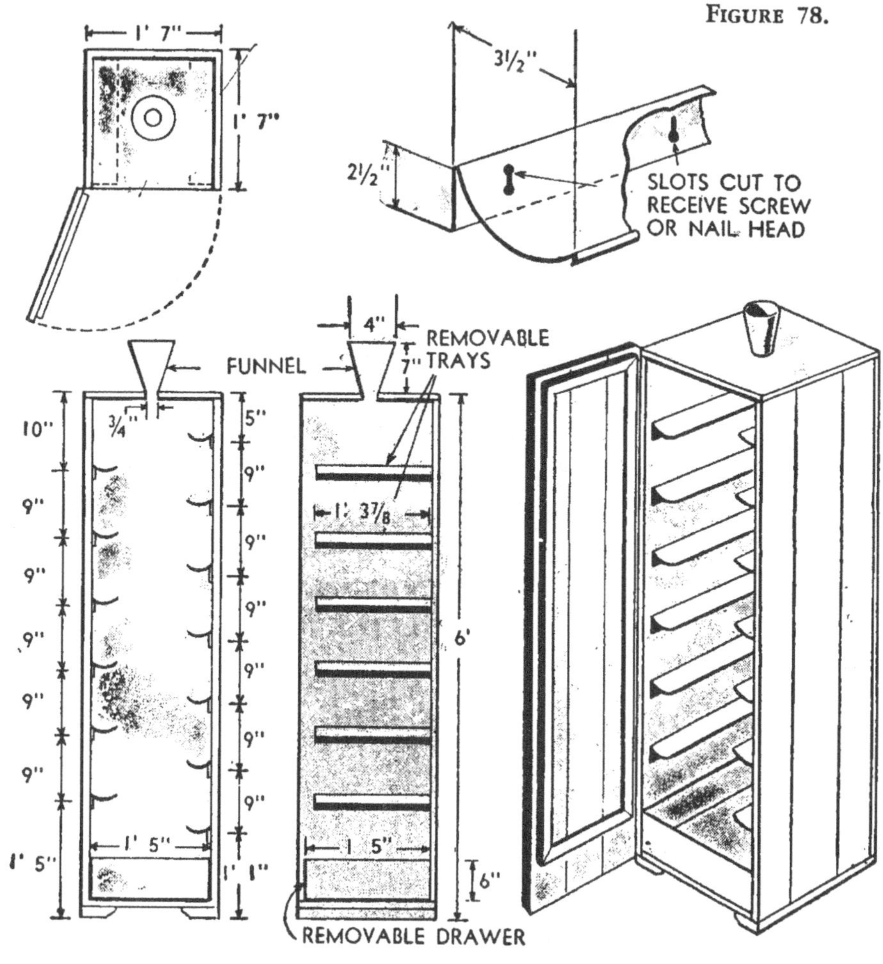

CHAPTER 5

PREFABRICATED MATERIALS

		Paragraphs
Section I.	Camouflage Clothing	61–62
II.	Net Sets	63–71

SECTION I. CAMOUFLAGE CLOTHING

61. GENERAL.—Prefabricated issue materials include camouflage clothing for individuals and net sets for weapons and aircraft. These issue materials are designed to solve special problems.

62. CAMOUFLAGE CLOTHING.—*a.* The snow suit (fig. 79) is a plain white, two-piece garment designed to blend with snow. The complete uniform for snow camouflage consists of the following items:

(1) Ski parkas and hood, reversible—white on one side, olive drab on the other.

(2) Ski trousers, white.

(3) White over-mittens.

FIGURE 79.—Mountain troops wearing snow uniform.

(4) Overcoat, parka type, pile lined, reversible—white on one side, olive drab on other.

(5) Mukluk boots.

b. The jungle suit, a reversible one-piece garment, mottled in shades of green on one side, the reverse side colored dark O.D., is no longer an item of issue. It has been replaced by an issue fatigue uniform colored dark O.D. No. 7.

SECTION II. NET SETS

63. NET SETS.—Net sets are designed to be carried conveniently and to be erected and dismantled quickly and easily.

64. NET SET, CAMOUFLAGE, NO. 1.—*a.* This net set was designed for the concealment of 3-inch antiaircraft artillery and is now replaced by net set No. 4.

b. The specification is: Corps of Engineers Tentative Specification T-1567.

65. NET SET, CAMOUFLAGE, NO 2.—*a.* This net set is organic equipment for concealment of light and medium field-artillery weapons. It provides a main flat-top, 29 feet square with a 14-foot slit embrasure, and 3 extensions, each 14 by 29 feet. All nets are garnished as one unit and are available in summer, winter, and desert colors.

b. The specification is: Corps of Engineers Tentative Specifications, T-1571.

66. NET SET, CAMOUFLAGE, NO 3.—*a.* This net set, which provides an embrasured flat-top, 29 feet square, is organic equipment for 37-mm and 40-mm antiaircraft artillery. It is issued in summer, winter, and desert colors.

b. The specification is: Corps of Engineers Tentative Specification T-1673.

67. NET SET, CAMOUFLAGE, NO. 4.—*a.* This net set, which provides an embrasured flat-top, 35 feet square, is organic equipment for 90-mm and 3-inch antiaircraft artillery. It is issued in summer, winter, and desert colors.

b. The specification is: Corps of Engineer Tentative Specification T-1674.

68. NET SET, CAMOUFLAGE, NO. 5.—*a.* This net set is organic equipment for concealment of 155-mm guns and 8-inch howitzers. It provides a flat-top, approximately 60 feet square, with a 30-foot embrasure. All nets are garnished as a unit and are issued in summer, winter, and desert colors.

b. The specification is: Corps of Engineer Tentative Specification T-1730.

69. NET SET, CAMOUFLAGE, NO. 6.—*a.* This net set is for single-engined fighter aircraft.

b. The specification is: Engineer Board Procurement No. 457.

70. NET SET, CAMOUFLAGE, NO. 7.—*a.* This net set is designed to cover a P-38 aircraft.

b. The specification is: Engineer Board Procurement No. 456.

71. NET SET, CAMOUFLAGE, NO 8.—*a.* This net set is designed for the concealment of 240-mm howitzers and 8-inch guns. It provides a flat-top, approximately 95 by 80 feet, with a 36-foot embrasure. All nets are garnished as a unit and are supplied in summer, winter, and desert colors.

b. The specification is not yet available.

CHAPTER 6

USE OF CLAY, PLASTER, PAPIER MÂCHÉ AND SIMILAR PLASTIC MATERIALS

		Paragraphs
Section I.	General	72
II.	Making Walls	73
III.	Methods of Moulding	74
IV.	Methods of Simulating Terrain Features.	75

SECTION I. GENERAL

72. GENERAL.—Plastic materials—such as plaster, lime, cement, plaster of Paris, native mud, clay, and many similar materials available commercially —are used in camouflage principally to construct decoys when it is necessary for them to possess three-dimensional form, a realistic appearance, and to be proof against close observation. Decoys made of plastic materials are employed in:

a. Operational camouflage, which includes construction of decoys representing military installations—for example, pill boxes or dragon's teeth (fig. 80).

b. Protective camouflage, which includes construction of, or additions to, civilian structures and terrain features such as hillsides and rock out-crops.

FIGURE 80.—Dragon's teeth of hollow plaster, a decov antitank obstacle.

SECTION II. MAKING WALLS

73. METHODS OF MAKING WALLS.—The choice of construction methods discussed below is determined by the weather prevailing in the area involved, the loads expected, and the height and resistance to wind required.

a. Solid walls by use of concrete forms.—(1) Construct a concrete wall form.

(2) Mix a sufficient quantity of clay and cement or mud and cement in the proportions of 7 parts of clay or mud to 1 part of cement. In relatively dry areas where there is enough sun to bake the wall, adobe made only of clay, sand, and water may be used. In other climates, adobe walls may be waterproofed by the application of bituminous emulsion to exposed surfaces. An approximately correct mixture for adobe is 10 cubic feet of a plastic clay, 10 cubic feet of sand, and 4 gallons of water.

(3) To this mixture add straw, grass, or other fibrous reinforcing material plus a few handfuls of lime, and mix thoroughly.

(4) If sandy earth is substituted for clay, change proportions to 3 parts of sand to 1 part of cement. In the absence of cement, use asphalt or similar adhesives.

(5) Pour resulting mixture into the form and puddle it so that no air pockets remain.

(6) Allow ample time for drying and remove form.

b. Light framework walls.—(1) Construct a framework of vertical and horizontal supports to the desired dimensions and cover with plain chicken wire or wooden laths.

(2) Mix moulding material as described above.

(3) Apply thickly by hand all over the wire or laths.

c. Walls of soil blocks.—(1) Mix molding materials as described above.

(2) Construct wooden moulds with a slight batter, 10 by 12 by 4 inches, 18 by 14 by 4 inches, or larger if necessary. It is also possible to make moulds by digging a level pit 4 inches deep in the ground, using greased or water-soaked boards for dividers.

(3) Pour mixture into the moulds and compact thoroughly with a tamper having about 8 square inches of tamping surface (fig. 81). Avoid the formation of layers.

(4) Allow to dry or "bake" in the sun.

(5) Remove moulds and allow blocks to harden until completely dry before using them as building blocks.

(6) To cement blocks together, use a mixture similar to original moulding mixture.

d. Bituminous stabilized soil blocks for wall construction.—

(1) The correct mixture must be determined by experimentation, as the ratio is affected by the type of clay available. One mix that has been used successfully is—

(*a*) 10 cubic feet of clay.
(*b*) 10 cubic feet of sand.
(*c*) 10 gallons of bituminous emulsion.
(*d*) 2 to 4 gallons of water.

(2) After drying approximately one day, the blocks are set on edge to facilitate further drying. Four weeks are required before the blocks are ready for use.

FIGURE 81.—Making soil blocks in wooden forms.

e. Sod blocks.—Sod blocks cut from heavy native sod may be of value in some situations.

f. Walls of burlap.—(1) Construct framework to required size and cover with burlap or other coarsely woven fabrics.

(2) Mix up plastic material to a consistency that will stick to the burlap but not run, and apply by hand.

(3) After walls have dried, they may be painted if desired.

SECTION III. METHODS OF MOULDING

74. MOULDING FORMS.—A rapid method of manufacturing complete objects or sections of complete objects is by moulds. Once a mould is prepared, a casting can be made in about 10 minutes. With two moulds working, 40 to 60 castings may be turned out in a day. Objects such as decoy land mines and rocks are a few of many which can be made in this way.

a. Materials for constructing mould.—

(1) Plaster of Paris (powdered gypsum).
(2) Animal or vegetable fibers.
(3) Grease or sterolene.
(4) Milled lumber or dressed timber.
(5) Nails, assorted.
(6) Burlap, osnaburg, or similar coarsely woven fabric.
(7) Paint for spraying.

b. Construction of a mould.—(1) Select the object which is to be cast. The maximum practical over-all size is 6 feet by 3 feet. Because of the weight, larger objects should be made in sections and the casts joined afterwards. Do not attempt to take a mould of an object which has undercuts or "lips" which will prevent the removal of the mould from the object.

(2) Coat object with oil, grease, or sterolene to minimize the chance of the mould sticking to the object. When taking a mould from a wall or any flat surface, enclose the area with a framework of light lumber (fig. 83).

(3) Mix up one bucket of plaster and water with three handfuls of fibrous material. Vary the constituents to obtain a consistency of thick cream. Pour a layer of the mixture into the mould and trowel it all over until it just covers the surface. It is necessary to work speedily in order to prevent the plaster hardening before the surface of the object is covered. One man should do the mixing and two men the trowelling. One and a half to two buckets of plaster cover an area 6 feet by 3 feet.

(4) Have burlap ready and cut to size. It should be slightly larger than the size of the mould so that all edges project a few inches. Dip it in water until it is saturated, then place it on the wet plaster. Press it on by patting with the hand until it adheres.

(5) Apply another layer of plaster.

(6) The plaster will have set sufficiently in 10 minutes for the frame to be removed. Some difficulty mav arise in removing the mould, but if the

FIGURE 82.—False plaster dragon's teeth.

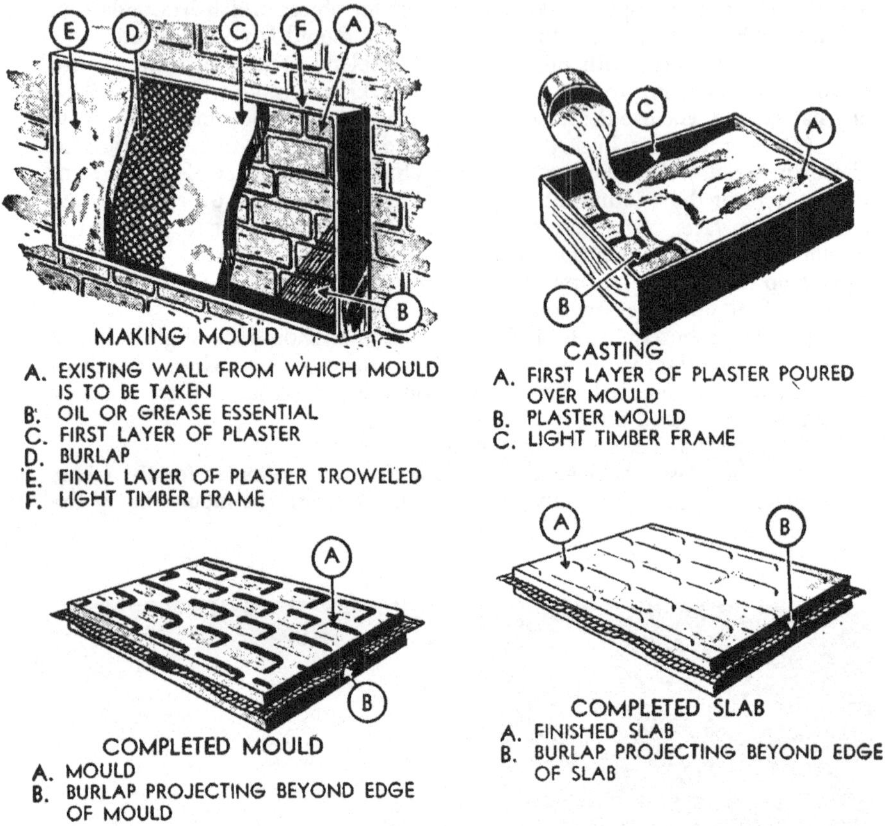

FIGURE 83.—Method of making a mould from an existing object and casting from the mould.

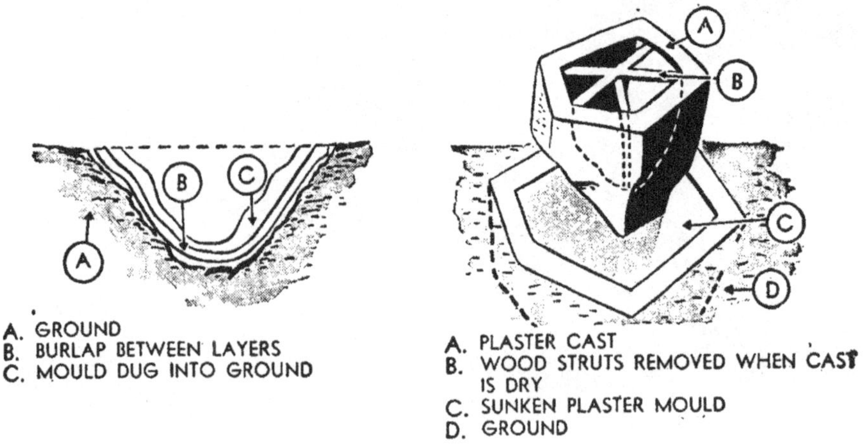

FIGURE 84.—Diagram showing method of making a mould in the ground.

edges are pried free with a chisel and pressure gradually brought to bear, the mould comes out readily.

(7) Depending on its type, the mould is prepared for use by placing it on a soft bed of burlap or by digging a hole and embedding the mould in the ground so that the exterior surface is supported (fig. 84).

(8) Paint the inside of the mould with oil paint. When dry, coat with grease, and the mould is ready for use.

c. Plaster casts from a mould.—(1) Coat mould thoroughly with oil, grease, or sterolene.

(2) Pour first layer of plaster and spread thinly, just covering the surface with a trowel.

(3) Have burlap ready and cut to size. The edges should project on all sides. Saturate it with water and place on the plaster, patting it well into the plaster.

(4) Pour in another layer of plaster and apply as before.

(5) After drying, the casting is removed and painted on both sides to waterproof it. It may then be textured and colored as desired.

d. Papier mâché casts from a mould.—This material requires more time than plaster, but is strong, durable, light in weight, and easily transportable.

(1) Make a smooth liquid paste of flour and water, the consistency of heavy cream. The water and flour may be boiled together; care should be taken to keep the paste from becoming lumpy.

(2) Tear up any available waste paper, thin cardboard, or fabric into 5- to 7-inch squares.

(3) Lightly oil or grease the mould.

(4) Saturate each piece of paper in the paste and then press it into the mould. Each piece is overlapped and pressed tightly to avoid air bubbles.

FIGURE 85.—A papier-mâché rock on a base of wire netting.

FIGURE 86.—Imitation rock formation made of plaster work.

(5) When one layer has been pressed into the mould, start the second and continue until 8 layers have been laid.

(6) Allow about one hour for drying and remove from the mould. Place in an oven or direct sunlight until thoroughly dried.

(7) Paint with shellac, varnish, or oleoresinous paint to waterproof the casting, then add any additional color coats by spraying until the desired effect is achieved.

Figure 85 illustrates a simulated rock made by applying papier mâché directly over a base of wire netting, without the use of a mould.

SECTION IV. METHODS OF SIMULATING TERRAIN FEATURES

75. SIMULATING TERRAIN FEATURES.—The following methods are applicable to temporary or semi-permanent installations in rear areas.

a. Simulated rock outcrops.—(1) Construct a reinforced skeleton framework, of the required height, to conform with the configuration of the adjacent area (fig. 86).

(2) Fasten a "sandwich" of chicken wire, burlap, and chicken wire to this framework.

(3) Mix up a plaster, without reinforcing material, of local mud, clay, or sand.

(4) Smear the mixture over the covered framework by hand, trowel, hoe, or improvised wooden spreader so that the final surface closely resembles the surface of real rocks in the neighborhood.

(5) Immediately imbed local rocks, gravel, and sand in the wet plaster and paint in whole or in part as required.

b. Simulated beach or sand dunes.—(1) Construct a framework to conform to the undulating appearance of these areas.

(2) Mix up a plaster.

(3) Dip strips of burlap of convenient size into the plaster mixture.

(4) Immediately place the burlap on the ground, plaster side down, and pat its entire surface so that the underside picks up the sand, pebbles, shells, and other inorganic material beneath it.

(5) Fasten burlap to framework and slope loose sand against the lower edges to blend the edges with the background.

CHAPTER 7

COVERAGE TABLES

76. NOTES ON COVERAGE TABLES.—The tables which follow give data on the coverage of various kinds of paints, stains, dyes, and adhesives appropriate to the fifteen different types of surfaces listed. The figures given are based upon gallons of unreduced paint and average painting conditions with operators of average experience. However, the rate of coverage and speed of application will vary with the condition of the surface to be coated, the equipment used, the skill of operators, and prevailing weather conditions. For example, on a large-scale disruptive painting operation using spray equipment the unit doing the work estimated that 50 percent of the paint consumed on the job was lost due to wind. For brushing, the figures are based on the use of a brush from 3 to 4 inches in width, unless otherwise stated under "Remarks." For spraying, figures are based on equipment meeting Specification No. T-1153. When equipment meeting Specification No. T-1291 is used instead, the coverage rate will remain approximately the same, but the spraying rate will be approximately 2½ times faster. The figures on speed of application are those of actual application and do not include time used in mixing paint or cleaning equipment. The figures on road distribution are calculated on the use of the bituminous distributor, model MX—described in TM 5-519-2—using an ⅛-inch nozzle and an 8-foot spraying bar with 2 men operating the distributor. For detailed information on preparation and application, refer to the appropriate table specified in the first column.

TABLE XLIX.—Asphalt, macadam, bituminous surfaces, and tar-coated roofing paper.

Coating Material	Brushing		Spraying		Asphalt Distributor		Remarks
	Coverage (square feet per gallon)	Man-hours (per gallon)	Coverage (square feet per gallon)	Man-minutes (per gallon)	Coverage (square feet per gallon)	Application Speed (square feet per minute)	
Paints							
Camouflage Oil-Type Runway Paint; Table VIII	300	3	350	7	350	28,000	FIRST CHOICE for traffic-bearing surfaces. Not applicable for roofing paper.
Camouflage Pigmented Bituminous Emulsions; Table VI	300	3	350	7	350	28,000	SECOND CHOICE for traffic-bearing surfaces. First choice for surfaces not subject to abrasion.
Oleoresinous Paint (Emulsifiable); Table II	450	4½	600	12	Not practical	Not practical	THIRD CHOICE for traffic-bearing surfaces. Second choice for surfaces not subject to traffic. Reduce with water.
Adhesives							
Bituminous Emulsion, Class A; Table XXVII	45	1/12	45	2	45	3,600	Brushed with large stable broom.
Cutback Asphalt; Table XXIX	30	1/12	30	2	30	2,400	Brushed with large stable broom.
Asphalt Cement; Table XXVIII	50	1/10	Not applicable	Not applicable	50	4,000	Brushed with large stable broom.

TABLE L.—Burlap, osnaburg, clothing, nets, and drapes.

Coating Material	Dipping		Spraying		Remarks
Paints	Coverage (square feet per gallon)	Man-minutes (per gallon)	Coverage (square feet per gallon)	Man-minutes (per gallon)	
Oleoresinous Paint (Emulsifiable); Table II	400	50	600	12	FIRST CHOICE. The figures for dip application are based on fabric impregnated by dipping and then passing through two steel squeeze rolls, 4 ft. long. The figures for spray application are based on coloring one side of the fabric only. Reduce paint with water.
Protein-Binder, Cold-Water Camouflage Paint; Table III	200	25	300	6	SECOND CHOICE. Same as above.
Field-Expedient, Emulsified Paint; Table XIX	40	5	60	2	THIRD CHOICE. Same as above.
Stains and Dyes					
Canvas Refinishing Compound; Table XVIII	Not applicable	Not applicable	Not applicable	Not applicable	May be used to re-color clothing. Application is by rubbing or impregnation by brushing. Not designed for other materials listed in heading.
Fabric Dye; Table XVII	Not applicable	Not applicable	Not applicable	Not applicable	For use on clothing only.
Adhesives					
Field-Expedient Green Vegetation Adhesive; Table XXX	Not applicable	Not applicable	Not applicable	Not applicable	Emergency use over small area such as truck tarpaulins.
Field-Expedient Starch Adhesive; Table XXXI	Not applicable	Not applicable	Not applicable	Not applicable	Same as above.

Table LI.—Canvas and duck.

Coating Material	Brushing		Rubbing		Remarks
	Coverage	Man-hours	Coverage	Man-minutes	
Stain Canvas Refinishing Compound; Table XVIII	Not available	Not available	Not available	Not available	Can be used on tentage, and on clothing. See also tables XV and XVI for expedient stains for tentage tonedown.

Table LII.—Feather-garnished and mineral-fiber-garnished netting.

Coating Material	Spraying		Remarks
	Coverage (square feet per gallon)	Man-minutes (per gallon)	
Paints			
Camouflage Pigmented Bituminous Emulsions; Table VI	300	4	FIRST CHOICE. Figures are based on coating one side of netting only.
Ready-Mixed, Oil-Type Camouflage Paint; Table I	350	4	SECOND CHOICE. Same as above.
Oleoresinous Paint (Emulsifiable); Table II	600	6	THIRD CHOICE. Same as above. Reduce with water.
Adhesives See table LIX. Glue-Paste Adhesive; Table XXXII	Not available	Not available	Emergency use if standard adhesives are not available for application of texturing materials to wire netting.

Table LIII.—Concrete, stucco, brick, building stone, slate.

Coating Material	Brushing		Spraying		Asphalt Distributor		Remarks
Paints	Coverage (square feet per gallon)	Man-hours (per gallon)	Coverage (square feet per gallon)	Man-minutes (per gallon)	Coverage (square feet per gallon)	Application Speed (square feet per minute)	
Camouflage Oil-Type Runway Paint; Table VIII	250	2½	300	6	300	24,000	FIRST CHOICE for traffic-bearing surfaces. Fourth choice for surfaces not subject to traffic.
Camouflage Pigmented Bituminous Emulsions; Table VI	250	2½	300	6	300	24,000	SECOND CHOICE for traffic-bearing surfaces. Third choice for surfaces not subject to traffic.
Oleoresinous Paint (Emulsifiable); Table II	400	4	500	10	Not practical	Not practical	THIRD CHOICE for traffic bearing surfaces. Fifth choice for surfaces not subject to traffic. Reduce with water.
Camouflage, Cement-Water, Paint, Type III; Table VII	300	3	350	7	350	28,000	FIRST CHOICE for surfaces not subject to traffic. Not practical for traffic-bearing surfaces.
Camouflage Cement-Water, Paint, Type I; Table VII	300	3	350	7	350	28,000	SECOND CHOICE for surfaces not subject to traffic. Not practical for traffic-bearing surfaces.
Field-Expedient, Emulsified Paint; Table XIX	50	½	50	2	50	4,000	Emergency use only.
Field-Expedient Cow-dung Paint; Table XXII	Not available	Not available	Not applicable	Not applicable	Not applicable	Not applicable	See table XXI.

TABLE LIII.—Concrete, stucco, brick, building stone, slate. ***Continued.***

Coating Material	Brushing		Spraying		Asphalt Distributor		Remarks
Stains	Coverage (square feet per gallon)	Man-hours (per gallon)	Coverage (square feet per gallon)	Man-minutes (per gallon)	Coverage (square feet per gallon)	Application Speed (square feet per minute)	
Cutback Asphalt, Types II, III; Table X	100	1	100	3	100	8,000	Expedient use only. Reduce with thinner to get desired color.
Used Crankcase Oil; Table XXIII	50	½	50	2	50	4,000	Expedient use only.
Xylene and Tar (RT–2); Table XIV	Not available	Not available	Not available	Not available	Not available	Not available	
Kerosene and Asphalt; Table XIV	Not available	Not available	Not available	Not available	Not available	Not available	
Adhesives							Brush with stable broom. A prime coat of the same adhesive should be applied prior to the finish coat at the following rates:
Bituminous Emulsion, Class A; Table XXVII	45	1/12	45	2	45	3,600	Reduce one part of adhesive to 3 parts of water—700 square feet per gallon.
Cutback Asphalt; Table XXIX	30	1/20	30	2	30	2,400	500 square feet per gallon.
Asphalt Cement; Table XXVIII	50	1/10	Not applicable	Not applicable	50	4,000	500 square feet per gallon.

TABLE LIV.—Earth, sand, and stabilized soil.

Coating Material	Spraying		Asphalt Distributor		Remarks
Paints	Coverage (square feet per gallon)	Man-minutes (per gallon)	Coverage (square feet per gallon)	Application Speed (square feet per minute)	
Camouflage Pigmented Bituminous Emulsions; Table VI	100	3	100	8,000	
Camouflage Bituminous-Emulsion Adhesive; Table IX	100	3	100	8,000	
Cutback Asphalt, Types I, II, III; Table X	100	3	100	8,000	Reduce with thinner to get desired color.
Field-Expedient Emulsified Paint; Table XIX	20	2	20	1,600	Emergency use only.
Stains					
Used Crankcase Oil; Table XXIII	20	2	20	1,600	
Raw Molasses; Table XXI	Not available	Not available	Not available	Not available	Emergency dust layer and tone-down, 5 to 20 square feet per gallon.

Table LV.—Glass.

Coating Material	Brushing		Spraying		Remarks
Paints	Coverage (square feet per gallon)	Man-hours (per gallon)	Coverage (square feet per gallon)	Man-minutes (per gallon)	
Ready-Mixed Oil-Type Camouflage Paint; Table I	500	5	600	12	FIRST CHOICE. If glass is set tightly in the frame, painting black is apt to cause cracking of the glass from the sun's heat.
Camouflage Pigmented Bituminous Emulsions; Table VI	400	4	475	10	SECOND CHOICE. Same as above.
Oleoresinous Paint (Emulsifiable); Table II	600	6	750	15	THIRD CHOICE. Same as above.
Gasoline-Soluble Camouflage Paint, Table IV	500	5	600	12	For temporary use, as for blackout purposes.
Adhesives					
Bituminous Emulsion, Class A; Table XXVII	45	½	45	2	
Cutback Asphalt; Table XXIX	30	½	30	2	
Field-Expedient Green Vegetation Adhesive; Table XXX	Not applicable	Not applicable	Not applicable	Not applicable	Emergency use.

TABLE LVI.—Glass fiber and steel wool.

Coating Material	Spraying		Remarks
Paints	Coverage (square feet per gallon)	Man-minutes (per gallon)	
Ready-Mixed, Oil Type, Camouflage Paint; Table I	350	7	FIRST CHOICE.
Oleoresinous Paint (Emulsifiable); Table II	600	12	SECOND CHOICE.
Enamel, Olive Drab, Rust-Inhibiting Army Specification 3–181	500	10	This is not a camouflage paint and is available in olive drab color only. Its use is recommended as a basic coating where humid atmospheric conditions without its application would shorten the life of the netting. The figures for spray application are based on coloring one side of the garnishing only.

TABLE LVII.—Grass.

Coating Material	Spraying		Asphalt Distributor		Remarks
Paints	Coverage (square feet per gallon)	Man-hours (per gallon)	Coverage (square feet per gallon)	Application Speed (square feet per minute)	
Camouflage Pigmented Bituminous Emulsions; Table VI	100	3	100	8,000	
Camouflage Bituminous Emulsion Adhesive; Table IX	100	3	100	8,000	
Cutback Asphalt; Table X	100	3	100	8,000	Reduce with thinner to get desired color.

TABLE LVIII.—Metal, already coated or primed.

Coating Material	Brushing		Spraying		Remarks
Paints	Coverage (square feet per gallon)	Man-hours (per gallon)	Coverage (square feet per gallon)	Man-minutes (per gallon)	
Lusterless Enamel; Table V	500	5	600	12	FIRST CHOICE.
Ready-Mixed Oil-Type Camouflage Paint; Table I	500	5	600	12	SECOND CHOICE.

Oleoresinous Paint (Emulsifiable); Table II	600	6	750	15	THIRD CHOICE.
Gasoline-Soluble Camouflage Paint; Table IV	500	5	600	12	For temporary use such as identification marks on vehicles, guns, aircraft, and equipment.
Camouflage Cement Water Paint, Type III; Table VII	450	4½	525	11	
Adhesives					
Bituminous Emulsion, Class A; Table XXVII	45	½	45	3	
Cutback Asphalt; Table XXIX	30	⅓	30	2	
Field-Expedient Green Vegetation Adhesive; Table XXX	Not applicable	Not applicable	Not applicable	Not applicable	Emergency use for small area such as on vehicles.
Field-Expedient Starch Adhesive; Table XXXI	Not applicable	Not applicable	Not applicable	Not applicable	As above.

Table LIX.—Metal, galvanized.

Coating Material	Brushing		Spraying		Remarks
Paints	Coverage (square feet per gallon)	Man-hours (per gallon)	Coverage (square feet per gallon)	Man-minutes (per gallon)	
Camouflage Cement Water Paint, Type III; Table VII	450	4½	525	11	FIRST CHOICE.
Camouflage Pigmented Bituminous Emulsions; Table VI	400	4	475	10	SECOND CHOICE.
Oleoresinous Paint (Emulsifiable); Table II	600	6	750	15	THIRD CHOICE.
Adhesives					
Bituminous Emulsion, Class A; Table XXVII	45	½	45	2	
Cutback Asphalt; Table XXIX	30	½	30	2	

TABLE LX.—Mineral-coated roofing.

Coating Material	Brushing		Spraying		Remarks
Paints	Coverage (square feet per gallon)	Man-hours (per gallon)	Coverage (square feet per gallon)	Man-minutes (per gallon)	
Camouflage Pigmented Bituminous Emulsions; Table VI	300	3	350	7	FIRST CHOICE.
Oleoresinous Paint (Emulsifiable); Table II	450	4½	600	12	SECOND CHOICE. Reduce with water.
Adhesives					
Bituminous Emulsion, Class A; Table XXVII	45	½	45	2	FIRST CHOICE.
Cutback Asphalt, Table XXIX	30	½	30	2	SECOND CHOICE.

TABLE LXI.—Rubber.

Coating Material	Brushing		Spraying		Remarks
Paints	Coverage (square feet per gallon)	Man-hours (per gallon)	Coverage (square feet per gallon)	Man-minutes (per gallon)	
Ready-Mixed, Oil-Type Camouflage Paint; Table I	400	4	475	10	FIRST CHOICE.
Oleoresinous Paint (Emulsifiable); Table II	450	4½	600	12	SECOND CHOICE.

TABLE LXII.—Wood.

Coating Material	Brushing		Spraying		Remarks
Paints	Coverage (square feet per gallon)	Man-hours (per gallon)	Coverage (square feet per gallon)	Man-minutes (per gallon)	
Ready-Mixed, Oil-Type Camouflage Paint; Table I	400	4	475	10	FIRST CHOICE. Unpainted surfaces should be primed with oil-base primer. Table XI.
Camouflage, Oil-Type Runway Paint; Table VIII	300	3	350	7	SECOND CHOICE.
Oleoresinous Paint (Emulsifiable); Table II	450	4½	600	12	THIRD CHOICE. Unpainted surface should be primed before application. For primer, see Table XI.
Camouflage Pigmented Bituminous Emulsions; Table VI	300	3	350	7	FOURTH CHOICE.
Field-Expedient Emulsified Paint; Table XIX	Not available	Not available	Not available	Not available	
Stains					
Creosote Stain, Ready-Mixed; Table XIII	200	2	200	4	Cannot be painted over with an oil paint in light colors because of the staining from creosote.
Oil Stain, Ready-Mixed; Table XIII	200	2	200	4	
Asphalt Stain, Ready-Mixed; Table XIII	200	2	200	4	
Adhesives					
Bituminous Emulsion, Class A; Table XXVII	45	½	45	2	

Cutback Asphalt; Table XXIX	30	½	30	2	
Field-Expedient Green Vegetation Adhesive; Table XXX	Not applicable	Not applicable	Not applicable	Not applicable	Emergency use on small areas.
Field-Expedient Starch Adhesive; Table XXXI	Not applicable	Not applicable	Not applicable	Not applicable	Emergency use.

TABLE LXIII.—Wood chips and other texturing materials.

Coating Material	Spraying		Asphalt Distributor		Remarks
Paints	Coverage (square feet per gallon)	Man-minutes (per gallon)	Coverage (square feet per gallon)	Application Speed (square feet per minute)	
Camouflage Pigmented Bituminous Emulsions; Table VI	150	3	150	12,000	FIRST CHOICE.
Camouflage Oil-Type Runway Paint; Table VIII	150	3	150	12,000	SECOND CHOICE.
Oleoresinous Paint (Emulsifiable); Table II	300	6	300	24,000	THIRD CHOICE.
Stains					
Creosote Stain, Ready-Mixed; Table XIII	200	4	200	16,000	
Oil Stain, Ready-Mixed; Table XIII	200	4	200	16,000	
Asphalt Stain, Ready-Mixed; Table XIII	200	4	200	16,000	

CHAPTER 8

SPECIFICATIONS
CAMOUFLAGE MATERIALS AND EQUIPMENT

		PARAGRAPHS
Section I.	Specifications Listed by Issuing Authority	78–82
II.	Specifications Listed by Subject.	83–94

77. NOTES ON SPECIFICATIONS.—In section I of this chapter specifications on camouflage materials are listed under the issuing authorities; in section II they are listed under subjects to which they apply. When specifications are revised, a letter is added to the specification number. For example, specification T-1212B indicates that specification T-1212 was superseded by T-1212A and that in turn by specification T-1212B. The proper method of ordering specifications, to be assured of obtaining the latest copies including any amendment thereto, is to request the latest issue of the specification by title and number only. The letter or letters preceding the number are considered part of the number. Copies of these specifications may be obtained from Office, Chief of Engineers, Washington 25, D. C.

SECTION I. SPECIFICATIONS LISTED BY ISSUING AUTHORITY

78. FEDERAL SPECIFICATIONS.—

DD-T-86 Tape; cotton, bleached, dyed or grey, general-use.
QQ-W-461 Wire, steel (carbon) ; bare and zinc-coated.
RR-R-571 Rope; wire.
SS-A-671 Asphalt, cut-back (for) road-work.
T-C-571 Cord; sash, cotton, braided.
T-T-881 Twine, cotton, seine.

79. U. S. ARMY SPECIFICATIONS.—

6-280 Cloth, cotton, osnaburg.
14-3 Marline, jute, tarred.

80. CORPS OF ENGINEERS, U. S. ARMY TENTATIVE SPECIFICATIONS.—

T-1080 Fence posts, drive type (barbed wire entanglement).
T-1103 Enamel, olive drab, glyceryl phthalate.
T-1153 Paint spraying unit (portable).
T-1183 Enamel, phenolic, olive drab, lusterless.
T-1212 Fabric, impregnated camouflage.
T-1215 Paint, camouflage, ready-mixed, oil type.
T-1219 Nets, camouflage, wire, fabric garnished.
T-1220 Nets, camouflage, twine, fabric garnished.
T-1224 Bituminous emulsions, camouflage.
T-1227 Paint, camouflage, gasoline-soluble.
T-1271 Paint, spraying equipment for air compressor unit.
T-1279 Paint, camouflage, oleoresinous (emulsifiable).
T-1284 Nets,[1] camouflage, wire, steel wool garnished.
T-1287 Paints, camouflage, primer for wood, oil type.
T-1288 Nets, camouflage, welded wire.
T-1291 Paint spraying unit, pressure type (mobile).
T-1359 Compressor, air, motorized (capacity 105 cubic feet per minute).

[1] Specification reads: "Netting."

T-1457 Pneumatic tools and accessories.
T-1481 Wire netting, camouflage.
T-1483 Maintenance manuals and spare parts catalogs for engineer equipment.
T-1534 Nets, camouflage.
T-1564 Machine, paint mixing, 50-gallon capacity, electric motor driven.
T-1565 Wringer, roll, rubber.
T-1566 Cutter, cloth, disk type, electric, portable.
T-1567 Net set, camouflage No. 1.
T-1571 Net set, camouflage No. 2.
T-1574 Wringer, roll, 12-inch with vat, portable.
T-1575 Pneumatic agitator.
T-1583 Paint spraying unit, atomization type (mobile).
T-1599 Enamel lusterless (olive drab, blue drab, gray, black.)
T-1669 Nets, camouflage, cotton, shrimp.
T-1673 Net set, camouflage No. 3.
T-1674 Net set, camouflage No. 4.
T-1705 Compressor, air, gasoline engine driven (capacity 16 cubic feet per minute).
T-1719 Nets,[1] camouflage, wire, glass fiber garnished.
T-1720 Grip, wire rope, and cable.
T-1730 Net set, camouflage No. 5.
T-1760 Enamel, lusterless (for coating metal).

81. MISCELLANEOUS SPECIFICATIONS.—

Quartermaster Corps; OQMG No. 3. Compound for treating tentage, canvas leggings, and duck.

82. ENGINEER BOARD PROCUREMENT SPECIFICATIONS.—

No. 147 Fabric, impregnated camouflage (substitute).
234 Paint, face-, camouflage.
244 Paint, camouflage, oil type, runway.
245 Paint, cement-water, camouflage.
257 Nets, camouflage (substitute).
296 Tank, paint mixing (50-gallon capacity).
386 Garlands, camouflage, prefabricated.
396 Fabric, impregnated camouflage flame resistant.
409 Nets,[1] camouflage, wire, feather garnished.
452 Strap, mounting, for field artillery camouflage net sets.
456 Camouflage net set No. 7, fighter-plane hide (P-38).
457 Camouflage net set No. 6, fighter-plane hide (single-engine type).

SECTION II. SPECIFICATIONS LISTED BY SUBJECT

83. ABBREVIATIONS USED IN THIS SECTION ARE:

EBP—Engineer Board Procurement Specification.
FS—Federal Specification.
OQMG—Office Quartermaster General Specification.
T—Corps of Engineers, U. S. Army Tentative Specification.
USA—U. S. Army Specification.

[1] Specification reads: "Netting."

84. PAINT.—

T-1184 General specification for finishing, treating, and painting.
T-1093 Paint, coldwater, protein binder, camouflage.
T-1103 Enamel, olive drab, glyceryl phthalate.
T-1183 Enamel, phenolic, olive drab, lusterless.
T-1215 Paint, camouflage, ready-mixed, oil type.
T-1224 Bituminous emulsions, camouflage.
T-1227 Paint, camouflage, gasoline-soluble.
T-1279 Paint, camouflage, oleoresinous (emulsifiable).
T-1287 Paints, camouflage, primer for wood, oil type.
T-1760 Enamel, lusterless (for coating metal).
T-1599 Enamel, lusterless (olive drab, blue drab, gray, black.)
OQMG No. 3. Compound for treating tentage, canvas leggings and duck.
EBP. No. 245. Paint, cement-water, camouflage.
EBP. No. 244 Paint, camouflage, oil type, runway.
T-1225 Phosphorescent coating system.
T-1213 Camouflage colors.
EBP. No. 234 Paint, face-, camouflage.
FS. SS-A-671 Asphalt, cut-back (for) road-work.

85. ADHESIVE.—

SS-A671 Asphalt, cut-back (for) road-work.
T-1224 Bituminous emulsions, camouflage.

86. PAINTING EQUIPMENT.—

T-1153 Paint spraying unit (portable).
T-1291 Paint spraying unit, pressure type (mobile).
T-1271 Paint spraying equipment for air compressor unit.
T-1359 Compressor, air motorized (capacity 105 cubic feet per minute).
T-1564 Machine, paint mixing, 50-gallon capacity, electric motor driven.
T-1575 Pneumatic agitator.
T-1583 Paint spraying unit, atomization type (mobile).
T-1457 Pneumatic tools and accessories.
T-1705 Compressor, air, gasoline engine driven (capacity 16 cubic feet per minute).
EBP. No. 296 Tank, paint mixing (50-gallon capacity).
T-1483 Maintenance manuals and spare parts catalogs for engineer equipment.

87. FABRIC.—T-1212 Fabric, impregnated, camouflage.

EBP. No. 396 Fabric, impregnated, camouflage, flame resistant.
EBP. No. 386 Garlands, camouflage, prefabricated.
T-1213 Camouflage, colors.
T-1452 Processing fabrics, threads and cordage for mildewproofing.
EBP. No. 147 Fabric, impregnated camouflage (substitute).

88. EQUIPMENT FOR CUTTING, PAINTING, AND TREATING FABRIC.—T-1566 Cutter, cloth, disk type, electric, portable.

T-1565 Wringer, roll, rubber.
T-1574 Wringer, roll, 12-inch with vat, portable.

89. NETTING.—
T-1219 Nets, camouflage, wire, fabric garnished.
I-1220 Nets, camouflage, twine, fabric garnished.
T-1284 Nets,[1] camouflage, wire, steel wool garnished.
T-1288 Nets, camouflage, welded wire.
T-1481 Wire netting, camouflage.
T-1534 Nets, camouflage.
T-1669 Nets, camouflage, cotton, shrimp.
T-1719 Nets,[1] camouflage, wire, glass fiber garnished.
EBP. No. 257 Nets, camouflage (substitute).
EBP. No. 409 Nets, camouflage, wire, feather garnished.

90. NET SETS.—
T-1567 Net set, camouflage, No. 1.
T-1571 Net set, camouflage, No. 2.
T-1673 Net set, camouflage, No. 3.
T-1674 Net set, camouflage, No. 4.
T-1730 Net set, camouflage, No. 5.
EBP. No. 456 Camouflage net set No. 7, fighter-plane hide (P-38).
EBP. No. 457 Camouflage net set No. 6, fighter-plane hide (single engine type).
EBP. No. 452 Strap, mounting, for field artillery camouflage net sets.

91. WIRE, ROPE, CABLE AND POSTS.—
FS. QQ-W-461 Wire, steel (carbon); bare and zinc-coated.
F.S. RR-R-571 Rope; wire.
F.S. DDD-T-86 Tape; cotton, bleached, dyed, or grey, general-use.
USA 14-3 Marline, jute, tarred.
T-1080 Fence posts, drive type (barred wire entanglement).
F.S. T-T-881 Twine, cotton, seine.
F.S. T-O-571 Cord; sash, cotton, braided.

92. SPECIAL DEVICES.—
EBP. No. 452 Strap, mounting, for field artillery camouflage net sets.
T-1720 Grip, wire rope, and cable.

93. GENERAL.—
USA 100-12 Method of test for infrared reflectance.
T-1213 Camouflage colors.
T-1483 Maintenance manuals and spare parts catalogs for engineer equipment.
T-1184 General specification for finishing, treating, and painting.

94. MISCELLANEOUS.—
T-1225 Phosphorescent coating system.
T-1452 Processing fabrics, threads, and cordage for mildew-proofing.
EBP. No. 234 Paint, face-, camouflage.
T-1720 Grip, wire rope, and cable.

[1] Specification reads: "Netting."

INDEX

	PARAGRAPHS	PAGE
Adhesives:		
Application of	8	29
Characteristics of issue	7*d*	29
Field expedient	9, tables XXX-XXXII	32, 34, 36
Methods of application of	8*b*	32
Preparation of surface for	8*a*	29
Selection of	7*c*	29
Sources of	7*b*, 9*b*	28, 32
Tables of	tables XXVII-XXXII	30, 31, 34-36
Uses of	7*a*	28
Adobe construction	73*a*	141
Aircraft concealment, prefabricated issue materials for:		
P-38	70	139
Single-engined aircraft	69	139
Aniline dye, use of, on turf	37*b* (3) (*d*)	94
Antiaircraft artillery, net sets for	66, 67	139
Apertures, screening of	17*c*	48
Artillery, net sets for	65-68, 71	139
Asphalt and kerosene stain for concrete	table XIV	20
Asphalt cement	table XXVIII	30
Asphalt stain for wood	table XIII	19
Asphalt surfaces, coverage on	table XLIX	148
Bamboo rope, making	27*b*	80
Birchbark,—a substitute for cloth	14*c*	44
Bituminous emulsions, use of:		
As adhesive (class A)	table XXVII	30
As paint:		
Pigmented (class B)	table VI	13
Unpigmented (class A)	table IX	15
As plastic material	73*d*	141
Cautions on use of	table VI	13
Bituminous surfaces, camouflage treatment of:		
Adhesive application	8, table XLIX	29, 148
Coverage on	table XLIX	148
Painting	tables VI, XLIX	13, 148
Blackout, treatment for	7*a* (2), table LV	28, 154
Blocks, methods of making soil	73*d*, 73*e*	141, 142

	PARAGRAPHS	PAGE
Brick surfaces coverage on	table LIII	151
British standard camouflage paints	tables XXIV, XXV, XXVI	27, 28
Burlap	table XXXIV, 21*a*, table XXXVIII	45, 65
Burlap surfaces, coverage on	table L	149
Cable	25*c*	76
Camcolour	table XXV	27
Camemulsion	table XXVI	28
Camouflage bituminous emulsion adhesive as paint	table IX	15
Camouflage cement-water paints	table VII	14
Camouflage clothing	62	138
Camouflage face paint	table XII	18
Camouflage materials, manufacture of, by factory methods in the field.	60	123
Applying chicken feathers to wire netting	60*g*	134
Dyeing troop clothing	60*e*	131
Garnishing twine nets	60*c*	124
Garnishing wire netting	60*b*	123
Making field-expedient cable	fig. 42, 25*d*	78, 79
Making garnishing strips	60*f*	131
Painting and cutting bolted fabric for garnishing	60*d*	125
Camouflage oil-type runway paint	table VIII	14-15
Camouflage pigmented bituminous emulsions	table VI	13
Canvas, coloring or recoloring, materials for:		
Canvas refinishing compound	tables XVIII, LI	23, 150
(*See also* Field-expedient tentage stains)		
Canvas refinishing compound	table XVIII	23
Casein paint (*See* Protein binder cold-water camouflage paint.)		
Cement-water paints, camouflage	table VII	14
Chemicals, use of, on turf:		
Coloring property of, general	6*c*	17
Dye, aniline	37*b* (3) (*d*)	94
Fertilizers	37*g* (2)	93
Killing agents	37*b*	94
Chen-chen, cloth substitute	14*c*	44
Cinders as texturing material	table XXXIII	41
Cloth (*See* Burlap, Osnaburg.)		
Clothing, camouflage, issue	62	138
Clothing, coverage of paints on	table I.	149
Cloth substitutes	14*c*	44
Clothing, troop, dyeing of:		
Fabric dye	table XVII	17
Factory method of	60*e*	131
Coal tar, as substitute for cutback asphalt	table X	16

	PARAGRAPHS	PAGE
Coconut fiber, shredded, as texturing material	table XXXIII	40
Coffee beans as texturing material	table XXXIII	40
Coir as garnishing material	23*c*	72
Coir (shredded coconut fiber) as texturing material	table XXXIII	40
Coloring (*See* surface to be colored.)		
Coloring cloth for garnishing and screening	60*d*	125
Coloring materials:		
Application of texture	6*e*	19
Chemicals	6*c*, 37*g*	17, 92
Dyes	6*b*, tables XVII, XVIII, 37*b* (3) (*d*)	16, 22-23, 94
Field expedients	6*d*, tables XIX-XXIII	17, 24-26
Paints	4, 5, tables I-XXIII	6, 8-26
Primers	6*a*	16
Stains	6*b*, tables XII-XVI	16, 18-21
Concrete surfaces, camouflage treatment of:		
Application of adhesives to	8	29
Coverage on	table LIII	151
Painting	table LIII	151
Staining	tables XIV, LIII	20, 151
Texturing, built-in	10*b*	38
Cottonseed hulls as texturing material	table XXXIII	40
Cover, overhead:		
Choosing cut foliage for	41, table XLIII	98, 101
Foliage preservatives	42	99
Hardwoods	41*c* (1) (*b*), table XLV	99, 103
Softwoods	41*c* (1) (*a*)	99
Tropical plants	41*c* (1) (*d*), table XLIV	99, 102
Flat-tops, methods of using natural materials on	40	96
Cut and tufted bunches of straw or grass	40*b*	96
Vines and tubers	40*d*	98
Wetted moss	40*e*	98
General	39	96
Transplanting large trees	45	112
General	45*a*	112
Materials and equipment, bill of	45*b*	
Procedure:		112
Digging	45*c*	112
Guying	45*g*	114
Loading	45*e*	113
Moving	45*d*	113
Planting	45*f*	114
Pruning	45*b*	114
Relative dimensions of balls to trunk	table XLVII	113
Transplanting small trees and shrubs:		
General	44	106
Materials and equipment, bill of	44*b*	107

	PARAGRAPHS	PAGE
Planting seasons	44*a*	106
Procedure for root protection	44*e*	109
Procedure in digging	44*c*, 44*d*	107
Procedure in planting balled trees	44*g*	110
Procedure in transplanting at site	44*f*	109
Pruning	44*h*	110
Transplanted trees:		
Care of	44*j*	111
Maintenance of	44*i*	111
Vines	43*a*	100
Propagation of	43*d*	102
Table of	table XLVI	104-105
Training	43*c*	102
Transplanting	43*b*	102
Coverage of chemical killing agents	37*b* (3)	94
Coverage of fertilizers	37*e* (6) (*c*), 37*g* (2) (*d*)	88, 94
Coverage of texturing materials	table XXXIII	40
Coverage of water on turf	37*c* (3)	85
Coverage tables for paints, stains, and adhesives according to surface to which applied	76, tables XLIX-LXIII	147, 148-161
Cowdung, as material for paint	table XXII	26
Crankcase oil, used as stain	table XXIII	26
Creosote stain for wood	table XIII	19
Cutback asphalt:		
As adhesive	table XXIX	31
As paint	table X	16
Debris, uses of:		
For ground patterning	38*a*	95
For overhead cover	23*e*, 40*c*	75, 96
Decoys, construction of:		
For operational camouflage	72*a*	140
For protective camouflage	72*b*	140
Moulding	74	142
Papier mâché	74*d*, fig. 85	145
Runway by burning	37*d*	85
Use of plastic materials for	72-75	140-146
Dirt. (*See* Earths, native)		
Ditches; simulation of:		
By mowing	37*f* (1)	90
Drapes:		
Coverage of	table L	149
Garnished twine nets	18*a*, table XXXVI	51
Small-mesh netting	15, table XXXV	47
Dyes:		
Aniline, for use on turf	37*b* (3) (*d*)	94
Canvas refinishing compound	table XVIII	23
Fabric	table XVII	22

	PARAGRAPHS	PAGE
Field expedient	*6b* (2)	16
General	*6b*	16
Natural materials used in preparation of	*6b* (2)	16
Earth, native:		
As pigment	*6d* (2)	17
Coloring properties of	4	6
Surface, coverage on	table LIV	153
Use of	30*d*	81
Enamel, lusterless	table V	12
Equipment, camouflage	46-60	115-137
General	46	115
Heavy	48	115
Miscellaneous	56	120
Rotary cloth cutter	58	122
Texture distributors	56	120
Wood-chipping machine	57	122
Painting equipment	49-55	115-120
Asphalt distributors	52	117
Commercial	53	120
Field-expedient paint sprayers	54	120
Mobile paint-spraying unit	51	116
Paint mixers	55	120
Portable paint-spraying unit	50	115
Simple tools	47	115
Excelsior as garnishing material	23*c*	72
Excelsior as texturing material	table XXXIII	40
Fabric, camouflage	14, table XXXIV	14, 45
Fabric dye	table XVII	22
Fabric-garnished netting	tables XXXVI-XXXVII	51, 55
Fabric scraps as texturing material	table XXXIII	40
Face and hands, camouflage treatment of:		
Face paint, camouflage	table XII	19
Feathers as garnishing material	table XXXVII, 23*d*	54, 72
Feathers as texturing material	table XXXIII	40
Feather surface, coverage on	table LII	150
Fertilizers	37*g* (2)	93
Field artillery, net sets for	65, 68, 71	139
Field crops, simulation of:		
By controlled irrigation	37*c*	85
By mowing	37*f* (2)	90
By use of chemical killing agents	37*b*	94
By use of fertilizer drills	37*g* (1) (*b*)	93
By use of fertilizers	37*g* (1)	92
Field expedients:		
(*See also* Camouflage materials, manufacture of.)		
(*See also* Dyes.)		
(*See also* Equipment, camouflage.)		

	PARAGRAPHS	PAGE
(*See also* Fabrics.)		
(*See also* Garnishing materials.)		
(*See also* Paint.)		
(*See also* Pigments.)		
(*See also* Screening materials.)		
(*See also* Stains.)		
(*See also* Structural materials.)		
(*See also* Tools and equipment.)		
Field-expedient adhesives	9, tables XXX-XXXII	32, 34-36
Field-expedient cowdung paint	table XXII	26
Field-expedient emulsified paint	table XIX	24
Field-expedient manufacturing devices:		
Feather machine	60*g*, table XLVIII	134, 135
General	59	122
Mat loom, construction of	16*a*, fig. 13	48
Paint sprayer	54	120
Pigment-collecting chamber	60*b* (5)	137
Pigment-manufacturing machine	60*b* (4)	135
Pulverizing rock for dry pigment	60*b* (3)	134
Tin-can cutter, construction of	60*b* (1), fig. 75	134
Field-expedient soap (for field expedient paint)	table XX	25
Field-expedient tentage stain No. 1	table XV	21
Field-expedient tentage stain No. 2	table XVI	21
Fine-mesh screening:		
As substitute for cloth	14*c*	44
For obscuring apertures	17*c*	50
For snow camouflage	17*c*	50
Flat-tops, garnishing of, by use of:		
Cut and tufted bunches of grass	40*b*	96
Debris	40*c*	96
Planting or transplanting vines	40*d*, 43	98, 100
Spreading wetted moss	40*e*	98
Using cut foliage	40*a*	96
Foliage, cut, approximate lasting qualities of:		
Table of hardwood	table XLV	103
Table of most common species	table XLIII	101
Table of tropical	table XLIV	102
Foliage preservatives, use of	42	99
Ammonium thiocyanate and sorbitol	42*b*	100
Calcium chloride	42*a*	99
Water	42*c*	100
Garnished netting	18	50
Garnishing material:		
Advantages of artificial	20*b*	65
Advantages of natural	20*c*, 23*f*	65, 75
Burlap	21*a*, table XXXVIII	65
Captured fabrics	21*e*	66

	PARAGRAPHS	PAGE
Cloth	21	65
Field preparation of:		
Debris	23*e*	75
Fabric:		
Coloring bolted fabric	21*g*, 60*d*	67, 125
Coloring precut rolls	21*g* (2)	67
Cutting garnishing fabric	21*g* (4)	69
Dipping	21*g* (1) (*a*)	67
G.I. can method	21*g* (1) (*b*)	67
Painting and cutting bolted fabric for garnishing	60*d*, 60*f*	125, 131
Feathers	23*d*, 60*g*	72
Application to twine nets	23*d* (4) *b*	74
Application to wire netting	23*d* (4) *a*	73
Assembly of feather-garnished wire netting for flat tops	23*d* (8)	75
Coloration	23*d* (5)	74
Coverage	23*d* (3)	73
Durability	23*d* (7)	74
General	23*d*	72
Quality of feathers to be procured	23*d* (2)	73
Troop manufacture of	60*f*	131
Uses	23*d* (6)	74
Glass fiber	23*a*, table XXXVII	72, 55
Locally manufactured fabrics	21*f*	66
Osnaburg	21*b*, table XXXIX	65, 66
Paper	23*b*	72
Prefabricated garlanded wire	21*c*	66
Processed wood fibers	23*c*	72
Salvaged fabrics	21*d*	66
Steel wool	22*a*, table XXXVII	71, 54
Tin cans	60*b*, 22*b*	134, 71
Garnishing twine nets	60*c*	124
Garnishing wire netting	60*b*	123
Gasoline-soluble camouflage paint	table IV	11
Glass-fiber garnished netting	table XXXVII, 23*a*	55, 72
Glass surface, camouflage treatment of:		
Application of adhesive	7*a* (2), table LV	28, 154
Painting	table LV	154
Glue-paste adhesive	table XXXII	36
Granules:		
Texturing material	table XXXIII	40
Grass (*See* Ground patterning.)		
Gravel as texturing material	table XXXIII	40
Green vegetation adhesive	table XXX	34
Greenhide rope	27*a*	79
Ground patterning:		
Burning dry grass	37*a*	85

	PARAGRAPHS	PAGE
Chemical killing agents	37*b*	94
Observations on use of	37*b* (2)	94
Suggested uses of	37*b* (1)	94
Types of	37*b* (3)	94
Fertilizing	37*g*	92
Effects possible by using	37*g* (1)	92
Recommended fertilizers and application	37*g* (2)	93
General	35	82
Grass planting, general	37*e*	86
Drainage	37*e* (4)	87
Fertilizing	37*e* (6)	87
Mulching	37*a* (3)	84
Planting grass, methods of	37*e* (7)	88
Seeding	37*e* (7) (*a*)	89
Sodding	37*e* (7) (*c*)	89
Sprigging	37*e* (7) (*b*)	89
Planting seasons	37*e* (2)	87
Preparation of seedbed	37*e* (5)	87
Selection of grasses	37*e* (1)	87
Soil suitable for	37*e* (3)	87
Ground painting, coverage	tables LIV, LVII	153, 156
Methods of	36	83
Mowing produces effective changes in	37*f*	90
Scarification	37*b*	84
Cultivating	37*b* (4)	84
Harrowing	37*b* (5)	84
Plowing	37*b* (2)	84
Rooting	37*b* (1)	84
Scarifying	37*b* (3)	84
Smoothing or blading	37*b* (6)	85
Spot and pattern irrigation	37*c*	85
Spreading material over a surface	37*a*	83
Observations on	37*a* (3)	84
Three-dimensional	38	95
Planting vines	43	100
Spotting movable material	38*a*	95
Transplanting bushes and shrubs	38*b*, 44	95, 106
Two-dimensional	37	83
Hedgerows, simulation of:		
By applying fertilizer	37*g* (1) (*a*)	92
By spotting tumbleweed or sagebrush	37*a* (3)	84
Individuals, camouflage of:		
Face paint	table XII	18
Kerosene and asphalt stain for concrete	table XIV	20
Leather chips as texturing material	table XXXIII	40
Leaves and chopped palm fronds as texturing material	table XXXIII	40
Legumes	32*a*	82

	PARAGRAPHS	PAGE
Lusterless enamel	table V	12
Macadam surfaces, coverage on	table XLIX	148
Manufacturing methods and apparatus (*See* Camouflage materials, manufacture of.)		
Marline	table XLII	79
Metal surfaces, camouflage treatment of:		
Coverage ot adhesives and paints on	tables LVIII, LIX	156, 158
Mineral-coated roofing:		
Coverage of paints and adhesives on	table LX	159
Molasses, raw:		
Used as dust preventative	table LIV	153
Used as paint	table XXI	25
Mosquito netting as substitute for shrimp net	15*c* (2)	47
Mud. (*See* Earths, native.)		
Mulching, as two-dimensional patterning method	37*a* (3)	84
Natural materials (*See also* Vegetation):		
Cover, overhead	Chap. 3, Sec. III	96
Dyes	6*b* (2)	16
General, introduction	29	81
Ground patterning	Chap. 3, Sec. II	82
Types of	30	81
Nets:		
Screening materials	15, table XXXV	44, 47
Supporting materials	tables XXXVI, XXXVII	50, 51, 54
Net sets:		
Prefabricated issue materials	Chap. 5	138
Oil stain for wood	table XIII	19
Oil-type camouflage paint primer for wood	table XI	18
Oil type camouflage paint, ready-mixed	table I	8
Oil-type camouflage paint, runway	table VIII	14-15
Oleoresinous paint (emulsifiable)	table II	9
Orchards, simulation of:		
By applying fertilizer	37*g* (1) (*d*)	93
Organization of manual	3	5
Osnaburg	table XXXIV, 21*b*, table XXXIX	45, 65, 66
Osnaburg surfaces, coverage on	table L	149
Paint	4, 5, 6, tables I-XXIII	6, 16, 8-26
Application, methods of. (*See* individual tables.)		
British	6*f*, tables XXIV-XXVI	19, 27-28
Causes for failure of	5*c*	6
Climatic limitations of particular. (*See* individual tables.)		
Colors available. (*See* individual tables.)		
Colors, table of standard camouflage	fig. 2	7
Coverage of, on particular surface. (*See* appropriate table, chap. 7.)		
Desirable characteristics of	5*d*	6
Drying time. (*See* individual table.)		
Durability of. (*See* individual table.)		

	PARAGRAPHS	PAGE
Equipment for	49-55	115-120
Face	table XII	18
Field expedient	6*d*	17
Binding mediums, sources of	6*d*	17
Field-expedient soap for	table XX	25
Manufacturing equipment for	60*b* (3)	134
Pigments for	6*d* (2)	17
Tables of	tables XIX-XXIII	24-26
General	4, 5	6
Infrared properties of camouflage	5*d*	6
Paint-making devices	60*b* (3)-(5)	134-137
Pigments, field expedient	6*d* (2)	17
Preparation of. (*See* individual tables.)		
Primers for	6*a*, table XI	16, 18
Removability of. (*See* individual table.)		
Selection of, general factors affecting	4, 5	6
Snow	5*f*	7
Sources of supply of	5*a*	6
Specifications	Chap. 8	162
Stock numbers. (*See* individual tables.)		
Substitutes for commercial issue	6	16
Surface, type of. (*See* individual tables.)		
Tables of particular	tables I-XXIII	8-26
Uses, recommended. (*See* individual tables.)		
Painting garnished nets	60*c* (4)	125
Paper, uses of:		
As garnishing	23*b*	72
As netting	15*c* (3)	47
As screening	14*c* (5)	44
As texturing	table XXXIII	41
For papier mâché	74*d*	145
Palm matting, substitute for cloth	14*c* (2)	44
Patch garnishing, materials for	21*d*	66
Pigments, field-expedient:		
General	6*d* (2)	17
Manufacturing equipment	60*b* (3)-(5)	134-137
Preparation of black	6*d* (3)	17
Pine bough tips as texturing material	table XXXIII	40
Pine cones as texturing material	table XXXIII	40
Planting seasons for transplanting trees, shrubs, and vines	44*a*	106
Plant mix, open graded aggregate for texturing	table XXXIII	40
Plastic materials, use of:		
General	72	140
Moulding forms	74	142
Construction of a mould	74*b*	142
Making papier-mâché casts from mould	74*d*	145
Making papier-mâche originals	fig. 85	145

	PARAGRAPHS	PAGE
Making plaster casts from mould	74*c*	145
Simulating terrain features	75	146
Rock outcrops	75*a*	146
Sand dunes on beach	75*b*	146
Walls, burlap, making of	73*f*	142
Walls, framework, making of	73*b*	141
Walls, soil blocks, making of	73*c*	141
Walls, solid, making of	73*a*	141
Prefabricated issue materials:		
Camouflage clothing	62	138
Jungle suit	62*b*	139
Snow suit	62*a*	138
General	61	138
Net set, camouflage, No. 1	64	139
Net set, camouflage, No. 2	65	139
Net set, camouflage, No. 3	66	139
Net set, camouflage, No. 4	67	139
Net set, camouflage, No. 5	68	139
Net set, camouflage, No. 6	69	139
Net set, camouflage, No. 7	70	139
Net set, camouflage, No. 8	71	139
Primers, paint:		
For wood	table XI	18
General	6*a*	16
Protein-binder, cold-water camouflage paint	table III	10
Purpose of manual	1	5
Quartermaster compound for tentage. (*See* Canvas refinishing compound.)		
Rawhide Rope. (*See* Greenhide rope.)		
Ready-mixed, oil-type camouflage paint	table I	8
Redwood fiber as texturing material	table XXXIII, 23*c*	41, 72
Repairing twine nets	19	56
Resins as source of adhesive	9*b* (4)	
Rock chips as texturing material	table XXXIII	40
Rock outcrops, simulation of. (*See* Plastic materials, use of.)		
Roofing paper as substitute for cloth	14*c*	44
Rope, field-expedient substitutes for	27	79
Rubber surface, coverage of paint on	table LXI	159
Runway paint, camouflage, oil-type	table VIII	14
Runway, simulation of:		
By burning dry grass	37*d*	85
Salvaged fabric as garnishing	21*d*	66
Sand fencing (*See* Snow fencing)	17*b*	50
Sawdust as texturing material	table XXXIII	41
Scarification	37*b*	84
Seaweed as source of adhesive	9*b* (3)	33
Seaweed as texturing material	table XXXIII	41
Scope of manual	2	5

	PARAGRAPHS	PAGE
Screen, field mat	16	48
Screening material:		
Cloth	14	44
Burlap	table XXXIV	45
Osnaburg	table XXXIV	45
Other sources of cloth	table XXXIV	45
Substitutes for	14*c*	44
Field-expedient mat screens	16	48
Construction of mat loom	16*a*, fig. 13	48
Method of weaving	16*b*	48
Fine-mesh screening	17*c*	50
General	13	44
Small-mesh screening	15, table XXXV	44, 47
Substitutes for shrimp nets	15*c*	46
Snow fencing used as	17*b*	50
Truck tarpaulins used as	17*a*	48
Screens, overhead:		
Simulating haystacks	40*b*	96
Using cut and tufted straw or grass	40*b*	96
Using cut trees and bushes	40*a*	96
Scrim netting	15*c*	46
Shredded paper as texturing material	table XXXIII	41
Skin, tonedown of	table XII	18
Slag as texturing material	table XXXIII	40
Slate surface, coverage on	table LIII	151
Snow fencing (*See* Screening materials.)	17*b*	50
Snow, materials for camouflage in:		
Face paint	table XII	18
Fine-mesh screening	17*c*	50
Paint	5*f*	7
Soap, field-expedient, for manufacture of paint	table XIX	24
Soil, stabilized. (*See* Earths, native.)		
Spanish moss	table XXXIII	41
Specifications, camouflage materials and equipment:		
Adhesives	85	164
Corps of Engineers, U. S. Army tentative	80	162
Engineer Board procurement	82	163
Equipment for cutting, painting, and treating fabric	88	164
Fabric	87	164
Federal	78	162
General	93	165
Miscellaneous	94	165
Miscellaneous agencies	81	163
Net sets	90	165
Netting	89	165
Paint	84	164
Painting equipment	86	164

	PARAGRAPHS	PAGE
Special devices	92	165
U. S. Army	79	162
Wire, rope, cable, and posts	91	165
Splices, wire	table XLI note	77
Spoil, toning down	table XXIII	26
Stabilization of earth surfaces, expedients for:		
Dust prevention using crankcase oil	table XXIII	26
Dust prevention using raw molasses	table LIV	153
Stains:		
Field-expedient	tables XV, XVI	21
For concrete:		
Asphalt and kerosene	table XIV	20
Xylene and tar	table XIV	20
For tentage:		
Field-expedient	tables XV, XVI	21
For wood surfaces:		
Asphalt type	table XIII	13
Creosote type	table XIII	13
Oil-type	table XIII	13
General	4, 6*b*	6, 16
Tables of	tables XII-XVI	18-21
Stakes	28	80
Starch adhesive, field-expedient	table XXXI	35
Steel-wool garnished netting	table XXXVII	54
Coverage on	table LVI	155
Stock numbers. (*See* appropriate table.)		
Stone surfaces, coverage on	table LIII	152
Straw mattting, substitute for cloth	14*c*	44
Structural materials:		
General	24	76
Posts, stakes, and other structural materials	28	80
Wire, cable, and field expedients, general	25*a*	76
Cable	25*c*	76
Fibrous supporting materials	26, table XLII	79
Field-expedient methods of making cable	25*d*, fig. 42	79, 78
Other materials	27	79
Wire	25*b*, tables XL, XLI	76, 77
Stucco surfaces, coverage on	table LIII	152
Substitutes. (*See* Field expedients.)		
Supporting materials:		
Netting, large-mesh, general	18	50
Twine nets, large-mesh	18*a*	50
Fabric garnished	table XXXVI	51
Garnishing with chicken feathers	23*d* (4) (*b*)	74
Ungarnished	table XXXVI	51
Repair of twine nets	19	56
Alteration of diagonally woven nets	19*f*	60

	PARAGRAPHS	PAGE
Field-expedient needles	19*b* (3)	56
General	19*a*	56
How to weave nets	fig. 27	60-64
Materials for	19*b*	56
Patching a large hole	19*e*	59
Repairing tears	19*c*	57
Weaving tears	19*d*	58
Wire netting, general	18*b*	52
Fabric garnished	table XXXVII	54
Feather garnished	table XXXVII	54
Glass-fiber garnished	table XXXVII	54
Steel-wool garnished	table XXXVII	54
Table of contents		1
Tampico, as garnishing material	23*c*	72
Tape	table XLII	79
Tar and xylene stain for concrete	table XIV	20
Tar-coated roofing-paper surfaces, coverage on	table XLIX	148
Tentage:		
Coloring of, with canvas refinishing compound	table XVIII	23
Field-expedient stains for	tables XV, XVI	21
Texturing materials:		
Abrasion, relative, of	table XXXIII	40
Application of	12	39
Characteristics of common	table XXXIII	40
Coloring of	11*b*, table LXIII	38, 161
Coverage of	table XXXIII	40
Durability of	table XXXIII	40
Equipment for distribution of	56	120
General	10	37
Light meter readings of	table XXXIII	40
Preparation of	11	38
Sources of	10*a*, table XXXIII	40
Table of	table XXXIII	40
Wood-chipping machine	57	122
Tin cans, uses of:		
As garnishing material	22*b*	71
As texturing material	table XXXIII	41
Preparation of	fig. 75, 60*b* (1)	134
Tools and equipment, general	46	115
Asphalt distributors	52	117
Commercial equipment	53	120
Field-expedient manufacturing devices, general	59	122
Camouflage factory methods in the field	60	123
Other field expedients for manufacture of camouflage materials	60*b*	134
Paint sprayers	54	120
Heavy equipment	48	115
Miscellaneous equipment	56-58	120-122
Distributers for texturing materials	56	120

	PARAGRAPHS	PAGE
Rotary cloth-cutter blade	58	122
Wood-chipping machine	57	122
Painting equipment	49	115
Mobile paint-spraying unit	51	116
Portable paint-spraying unit	50	115
Paint mixers	55	120
Simple tools	47	115
Tracks, obscuring of:		
By scarification	37*b*	84
Trees:		
Planting of. (*See* Cover, overhead.)		
Simulation of	37*g* (1) (*d*)	93
Truck garden, simulation of:		
By mowing	37*f* (2)	90
By using tumbleweed or sagebrush	37*a* (1)	84
Turf. (*See* Ground patterning.)		
Used crankcase oil. (*See* Crankcase oil, used.)		
Vegetation:		
(*See also* Cover, overhead)		
(*See also* Ground patterning)		
As source of adhesive	table XXX	34
Desirable characteristics of	31	81
Limitations of	34	82
Types of	32	82
Uses of	33	82
Vines, table of	table XLVI	104
Walls, methods of constructing false	73	141
Weapons, concealment of, prefabricated issue materials for:		
Field artillery, light and medium	65	139
37-mm and 40-mm antiaircraft artillery	66	139
90-mm and 3-inch antiaircraft artillery	67	139
155-mm guns and 8-inch howitzers	68	139
240-mm howitzers and 8-inch guns	71	139
Wire	25, tables XL, XLI	76, 77
Wire, prefabricated, garlanded	21*c*	66
Woodchips as texturing material	table XXXIII	40
Coverage of paints and stains on	table LXIII	161
Wood cubes as texturing material	table XXXIII	40
Wood fibers, processed	23*c*	72
Wood stains	table XIII	19
Wood surfaces, camouflage treatment of:		
Coverage of adhesives, paints, and stains on	table LXII	160
Wood surfaces, primer for	table XI	18
Xylene and tar stain for concrete	table XIV	20

Made in the USA
Monee, IL
08 November 2026

67383092R00111